City Building on the Eastern Frontier

CREATING THE NORTH AMERICAN LANDSCAPE

Gregory Conniff
Edward K. Muller
David Schuyler
Consulting Editors

George F. Thompson
Series Founder and Director

Published in cooperation with the Center for American Places, Santa Fe, New Mexico, and Staunton, Virginia

City Building on the Eastern Frontier

Sorting the New Nineteenth-Century City

DIANE SHAW

The Johns Hopkins University Press
Baltimore and London

© 2004 The Johns Hopkins University Press
All rights reserved. Published 2004
Printed in the United States of America on acid-free paper
9 8 7 6 5 4 3 2 1

The Johns Hopkins University Press
2715 North Charles Street
Baltimore, Maryland 21218-4363
www.press.jhu.edu

Library of Congress Cataloging-in-Publication Data

Shaw, Diane, 1959–
   City building on the eastern frontier : sorting the new nineteenth-century city / Diane Shaw.
      p.   cm. — (Creating the North American landscape)
   Includes bibliographical references and index.
   ISBN 0-8018-7925-6 (hardcover : alk. paper)
   1. City planning—New York (State)—History—19th century.   2. Cities and towns—New York (State)—History—19th century.   3. Architecture—New York (State)—History—19th century.   I. Title.   II. Series.
   HT167.N5S53 2004
   307.1′216′0974709034—dc22   2003024909

A catalog record for this book is available from the British Library.

Pages 211–13 constitute an extension of this copyright page.

Family

Contents

Acknowledgments ix

ONE
Vernacular Urbanism and the Mercantile Network of New Cities 1

TWO
Planning the Sorted City:
Commercial, Industrial, and Civic Districts 19

THREE
Building the Sorted City: *The Three Epitome Districts* 41

FOUR
Refining the Sorted City: *Appearances in the Commercial District* 64

FIVE
Gentrifying the Sorted City:
Social Sorting in the Commercial District 87

SIX
The Reynolds Arcade and Athenaeum 123

SEVEN
Transportation and the Changing Streetscape 139

Notes 157

Index 201

Illustrations precede page 1.

Acknowledgments

This project began as something of an adventure. With an 1841 illustrated gazetteer of New York State cities in one hand and a modern AAA road map in the other, I crisscrossed the state, looking for signs of the nineteenth century. Of course, I had anticipated that the passage of time, not to mention the ravages of urban renewal, would be an obstacle, but there were times I simply could not correlate what I was seeing with the historical views. The Erie Canal was notoriously hard to find, having been largely paved over, although appellations such as Erie Boulevard in Syracuse provided some historical memory of the original artery. Even better was Rochester's Broad Street bridge, which was carried across the Genesee River on the original stone arches of the Erie Canal boat aqueduct. There were other persisting elements too. Next to a cliff of industrial buildings, the Genesee still plunged into a dramatic waterfall. The Four Corners was still a prime commercial intersection, which even included a 1920s replacement of the 1820s arcade. And a nearby block still held a Monroe County courthouse and other civic institutions. The original pattern of new cities sorted into commercial, industrial, and civic districts proved durable. Perhaps I could find the nineteenth century after all.

I could not have truly found it, however, without help. As historians, we are indebted to those whose research and writing has influenced our own work. I trust that this book and its citations indicate where many of my intellectual debts lie. But every author is further indebted to the people and institutions that helped bring the book to fruition. It is with pleasure and gratitude that I can try to thank them here.

A number of institutions provided valuable financial support. A University Fellowship and Humanities Graduate Research Grant from the University of California at Berkeley provided a year's stipend to support my initial research. An Albert J. Beveridge Grant from the American Historical Association, a Carter Manny Citation of Merit from the Graham Foundation, and a Berkman Faculty Development Grant from Carnegie Mellon University (CMU) helped me shape the project. A subsequent fellowship from the New York State Archives Larry J. Hackman Residency Program enabled me to do additional research. A summer stipend from the late Paul Christiano, then provost at Carnegie Mellon, and a grant from the CMU-administered Falk Fund provided additional research support to complete the book. Departmental support from the CMU School of Architecture allowed me to present portions of this work at the Vernacular Architecture Forum, Society of Architectural Historians, Society of American City and Regional Planning History, Association of American Geographers, and New York State History conferences.

Librarians, archivists, and curators have provided particular assistance in this project, especially Judy Haven and Michael Martin of the Onondaga Historical Association, Karl Kabelac and Melissa Mead of the Department of Rare Books and Special Collections at the University of Rochester Libraries, Andrew Kitzmann of the Erie Canal Museum, Jane Mebus at the Philadelphia Print Shop, and Craig Williams of the New York State Museum. The staff at the Rochester Pubic Library, the New York Public Library, and the Columbia University libraries were extremely helpful. The Inter-Library Loan staff at Carnegie Mellon did a superb job.

I remain indebted to my dissertation advisors Dell Upton, Paul Growth, and Mary P. Ryan, who challenged me to think about the multiple dimensions of nineteenth-century urbanization. Since that time this project has profited from the careful readings that friends and colleagues have given the manuscript. Ted Muller and Elizabeth Thomas deserve tremendous credit for helping me navigate between the big picture and the local story of these new cities. Members of the Vernacular Architecture Forum and a National Endowment for the Humanities summer institute provided valuable and collegial discussion. Under real time pressure Bill Littmann gave the final manuscript his scholarly and editorial review. Martha McNamara has offered unflagging moral support. Maureen Fellows and Brian Ladd even offered their homes. Sergio de Orbeta provided great technical assistance. Elizabeth Gratch copyedited the manuscript with a thoughtful hand. George Thompson and Randy Jones of the Center for American Places have been true and patient backers of the project.

This book has turned out to be a family affair. My sister edited an early version with her characteristic grace, humor, and eagle eye. My father, a model parent-professor, gave a thoughtful and nuanced reading of a later version. My mother and brother both sensibly knew when not to ask about

ACKNOWLEDGMENTS   xi

the book's progress. My husband, Kai Gutschow, deserves even more credit for this project than his splendid maps suggest. His scholarship and designer's eye helped shape this book. He has been my partner and champion through the entire process. To my children, Josefina and Benjamin, who only know that I have been busy, I say, whew, let's play.

City Building on the
Eastern Frontier

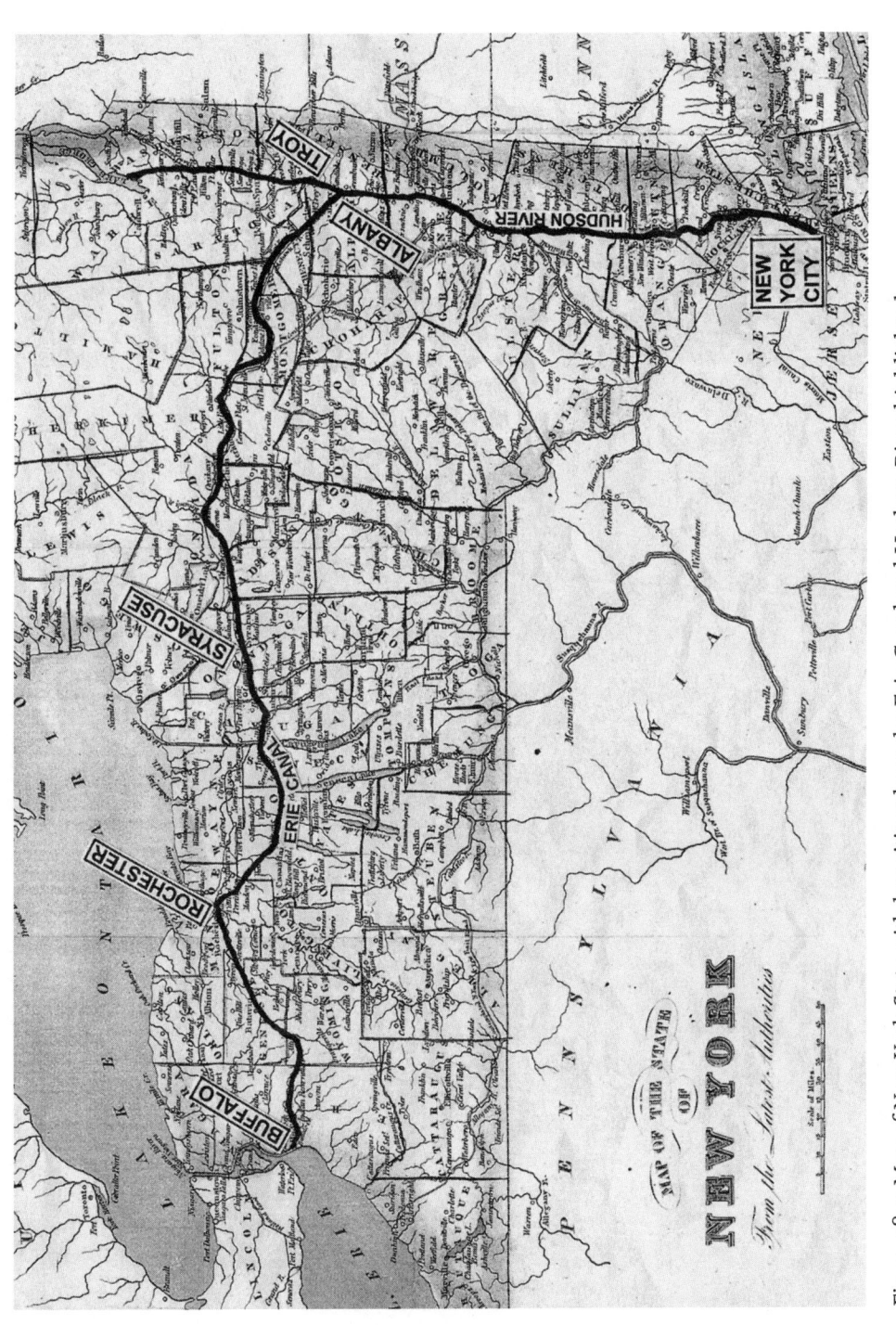

Fig. 1. 1841 Map of New York State with key cities along the Erie Canal and Hudson River highlighted.

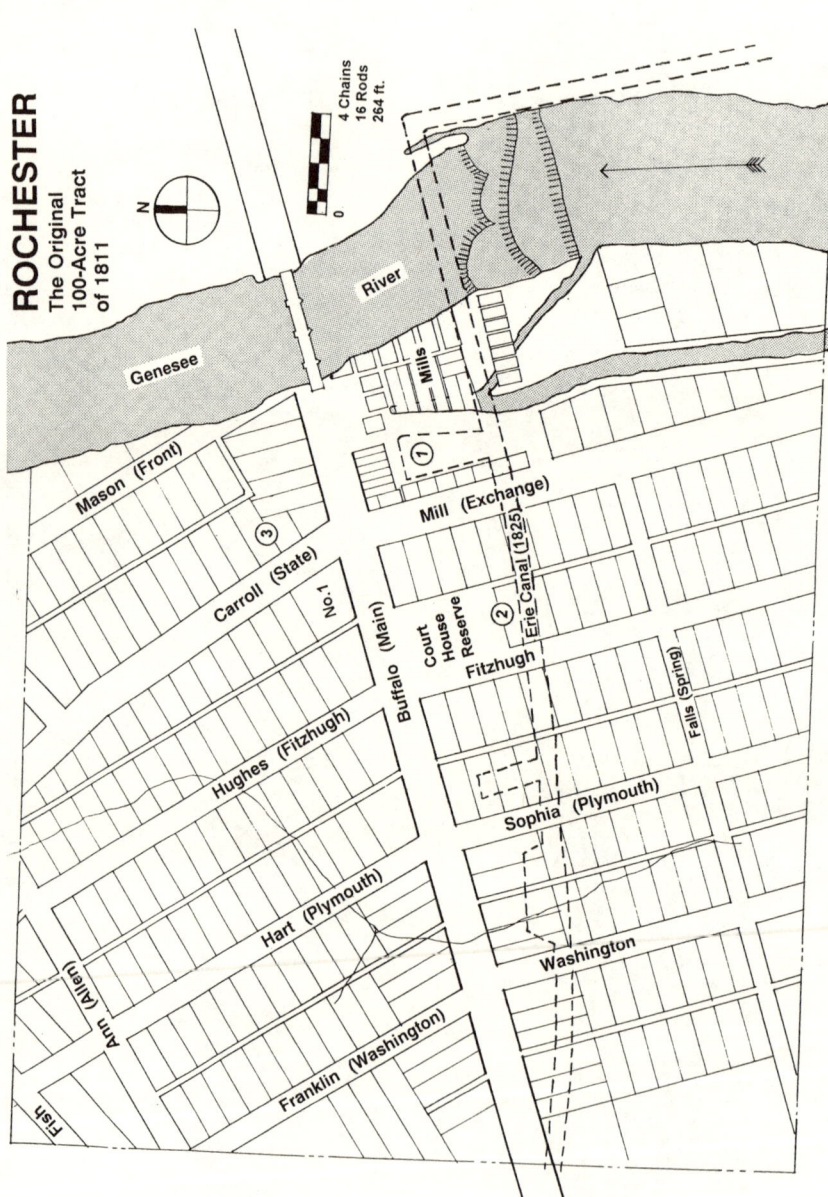

Fig. 2. Redrawn 1811 Plat of Rochester. Within the Four Corners formed by the intersection of Buffalo, Mill, and Carroll streets, Nathaniel Rochester sorted commercial, milling, and civic functions. At the 1825 opening of the Erie Canal, the procession encompassed: (1) The canal turnout at Child's Basin, (2) the Presbyterian church behind the courthouse, (3) the vicinity of Christopher's Mansion House hotel.

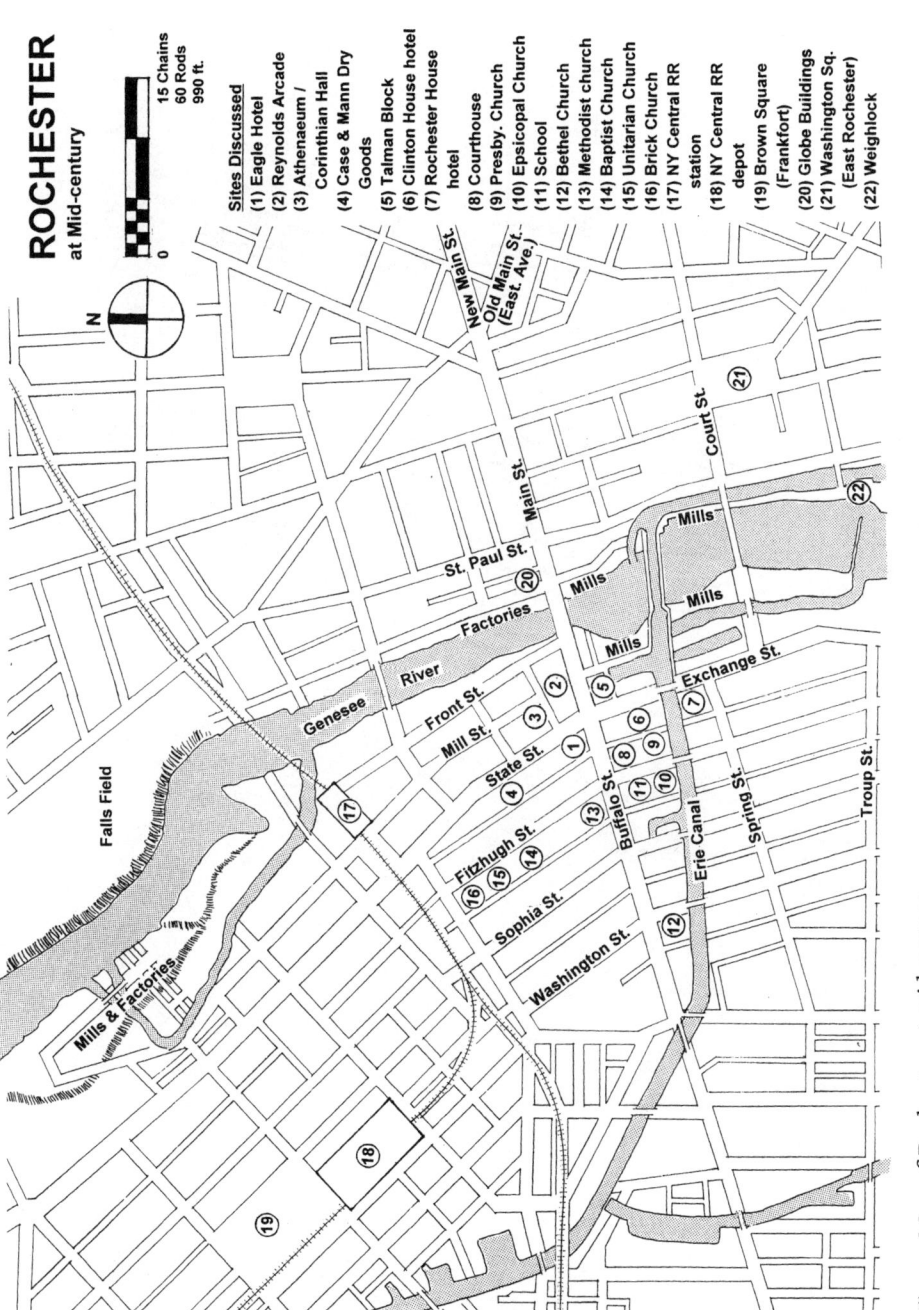

Fig. 3. Map of Rochester at midcentury.

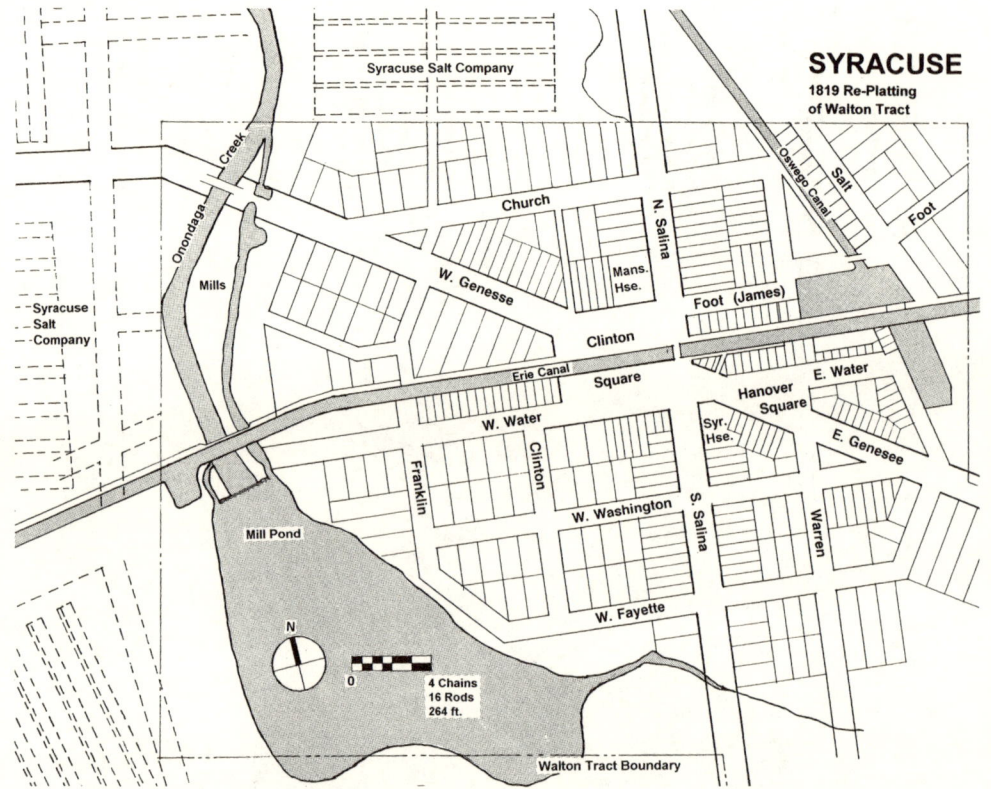

Fig. 4. Redrawn 1819 Plat of Syracuse. The juncture of Salina Street, the Genesee Turnpike, and the Erie Canal formed the city center at Clinton Square and the adjacent triangular Hanover Square.

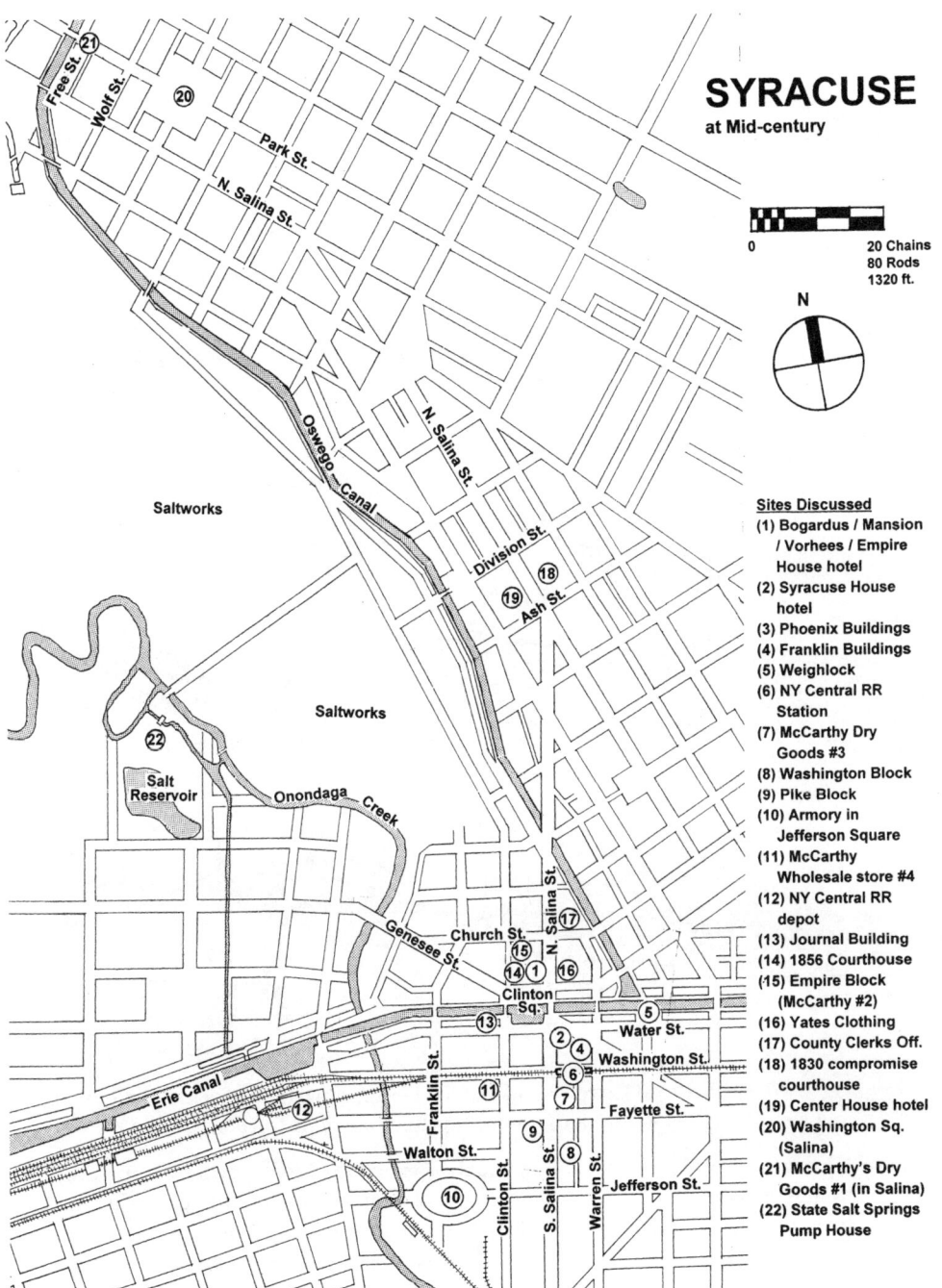

Fig. 5. Map of Syracuse at midcentury.

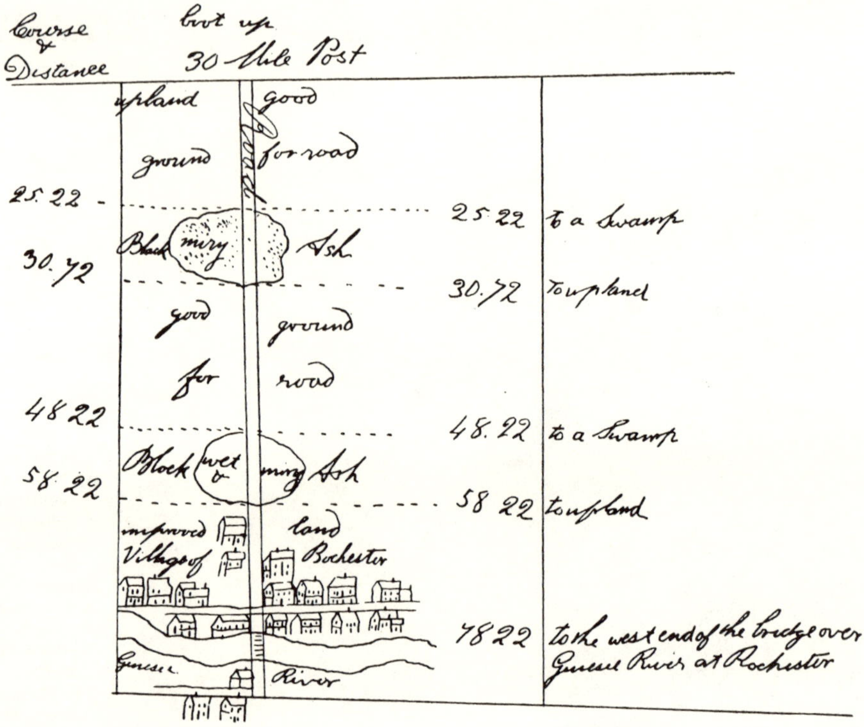

Fig. 6. (*top*) 1815 road surveyors map of Rochester, showing that actual settlement followed Nathaniel Rochester's compact plan at the Four Corners.

Fig. 7. (*bottom*) Sketch of Clinton Square during the 1820s, drawn from memory by a Syracuse resident. A contiguous row of shop houses and the Bogardus Inn (later the Mansion House hotel) defined the northern edge of the compactly built Clinton Square. Other than their size, little distinguished these commercial buildings from houses.

Fig. 8. (*top*) Basil Hall's 1827 Sketch of Buffalo Street, Rochester. Architectural distinctions clearly delineate the adjacent commercial and civic districts. Shop houses feature display windows and signboards. The civic "pocket" features the high-style Monroe County Courthouse, behind it the Presbyterian church; across the street peeks the Episcopal church tower and, in front of the courthouse terrace, stands a law office.

Fig. 9. (*bottom*) Duttenhofer's 1826 Sketch of Erie Canal Aqueduct, Rochester, showing the proxemics of milling and shipping. The canal aqueduct crosses above the Genesee River, passes the mills, and enters the canal basin within the industrial district.

Fig. 10. (*top*) 1837 view of the maturing Rochester civic pocket at Buffalo and Fitzhugh streets. The Monroe County Courthouse is encircled by a pair of temple-fronted law and medical offices, a Neoclassical church with a small chapel or Sunday school beside it, a Gothic Revival church, and an adjacent Greek Revival public school. A fence and landscaping complete the elegant tableau.

Fig. 11. (*bottom*) Photograph of the north side of Clinton Square in Syracuse, c. 1890. The Italianate Onondaga Courthouse (1857) forms a small, but architecturally distinct, civic pocket on the edge of the commercial space dominated by the Empire Hotel and Block (1847), built upon the original Bogardus Inn site.

Fig. 12. (*top*) 1838 view of a merchant milling operation at Child's Basin, Rochester. The millrace powered the mills, and the Erie Canal transported the flour to distant markets on waiting canal boats.

Fig. 13. (*bottom*) 1838 view of the Rochester House hotel facing Child's Basin on Exchange Street. This stylish hotel functioned as a merchants' exchange where businessmen gathered. It marks the boundary between the blue-collar and white-collar worlds in the southeast of the Four Corners.

Fig. 14. 1838 view of the Middle Falls of the Genesee River, by W. H. Bartlett. With its superior waterpower, the Frankfort tract's Middle Falls became Rochester's primary industrial district. The natural sublime of the dramatic falls mingles with the industrial sublime of the cliff-side mills and factories. The observation tower of Reynolds Arcade rises from the commercial district in the background.

Fig. 15. 1860 view of the variety of salt manufactories in Syracuse. Solar evaporation vats have movable lids in case of rain. Chimneyed sheds enclose the wood-fueled, manually stoked salt boiling blocks. Salt blocks line the Oswego feeder canal linking Salina to Syracuse. Top-hatted and top-coated gentlemen survey several of the scenes with interest.

Fig. 16. Barber and Howe's 1841 view of Buffalo Street, Rochester, looking eastward past the Methodist church (*left*), National Hotel (*left*) and courthouse (*right*), beyond the Four Corners, and over the Buffalo Street bridge to the east side. The image of sorted commercial and civic districts persist in this tidy streetscape, whose very awning posts are regulated by municipal ordinance.

Fig. 17. Barber and Howe's 1841 view of Clinton Square, Syracuse. The view looks eastward across Clinton Square and obliquely into Hanover Square. The Erie Canal flows under the Salina Street bridge. The double-fronted Coffin Block and attached Phoenix Buildings form a line of warehouses on their canal and shop fronts on their Hanover Square side. The balconied Syracuse House hotel figures prominently on the corner. Urban life is idealized as orderly.

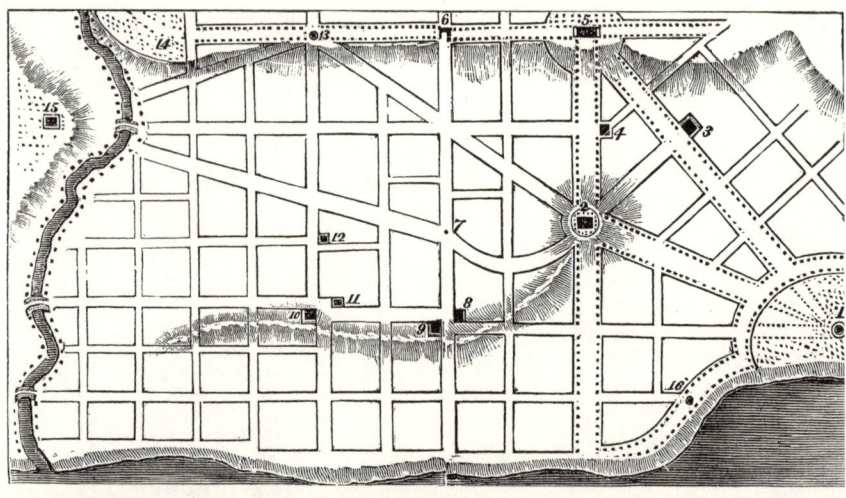

Fig. 18. (*top*) 1830 diagram of an ideal city, one critic's response to American city building practices. In his vision the entire urban composition is designed with an eye toward beauty rather than functionality.

Fig. 19. (*bottom*) The Syracuse House hotel, completed 1822. Booster Joshua Forman was "anxious that we should put up the best hotel west of Albany, as he thought it would be an inducement to others to purchase lots and start a village."

Fig. 20. (*top*) The rebuilt Syracuse House hotel, completed 1829. One resident recalled that "it was like a bouquet in a mud-hole." The street level veranda provided a prominent place for men to congregate, while the upper balconies removed women from public scrutiny.

Fig. 21. (*bottom*) The Syracuse House hotel and upgraded business rows on South Salina Street in 1856. The landmark hotel remained a popular gathering place at the juncture of Clinton and Hanover squares.

Fig. 22. (*top*) 1850s photograph of Clinton Square looking east. The buildings, if not the utterly tidy landscape, correspond closely to the engraving by Barber and Howe a decade earlier (Fig. 17), including the Daily Star Building on the north side of the canal and the Coffin Block and contiguous Phoenix Buildings on the south side. The Syracuse House hotel stands at the far right.

Fig. 23. (*bottom*) Nineteenth-century photograph of the Greek Revival Phoenix Buildings, constructed after the 1834 fire on Hanover Square, Syracuse. Taking their cues from the Syracuse House hotel, this rebuilding in turn precipitated similar architectural improvements in the facing Franklin Buildings across the square.

Fig. 24. (*top*) 1850s photograph of Clinton Square looking north, where the Salina Street bridge crosses the Erie Canal. The high-style 1847 Empire House hotel (*left*) contrasts sharply with the older gable-roofed Daily Star Building across the street. The attached Empire Block stretched along North Salina, the seat of dry goods row by midcentury. With its elegant cupola and interior rotunda, the Empire House challenged the Syracuse House diagonally across the canal.

Fig. 25. (*bottom*) 1867 photograph of double-fronted buildings along the canal. The Coffin Block has been razed for a new bank under construction. The commercial fronts of the Phoenix Buildings are visible facing Hanover Square. Opposite the canal, business rows display their warehouse fronts to the towpath.

Fig. 26. (*top*) Nineteenth-century photograph of canal warehouses and the Syracuse weighlock to the east of Clinton and Hanover squares. Chutes projecting from the loading bays discharge warehoused goods into canal boats. The planked towpath serves as both a loading dock and a place for male sociability. The Doric pillars of the weighlock provide a classical arcade that was both a functional shelter for the boats and an architecturally symbolic statement of the authority of the state.

Fig. 27. (*bottom*) 1890s photograph of the triple-fronted Journal Building, southwest side of Clinton Square, Syracuse. The canal front is plainest, reflecting its utilitarian warehousing activities. The landing front is pilastered, reflecting its prominent location within a central location. Yet even this facade is bifurcated, alluding to its continuation of the presumably fanciest Water Street retail facade, whose stringcourse wraps partially around the corner.

Fig. 28. (*top*) 1890 photograph of the east side of South Salina Street between Fayette and Jefferson streets showing the southward march of business and the coterminous rise of the Italianate style in Syracuse. Dillaye's trendsetting Washington Block rises in the middle of the block.

Fig. 29. (*bottom*) 1878 view of the Italianate Pike Block (1850), southwest corner of Salina and Fayette streets, Syracuse. The business block's large size and freestanding form characterized the new scale of investment in commercial enterprises. Its location marked the southward trend of business in town. The Bryant & Stratton Business College in the top story replaced an earlier business college in the building.

Fig. 30. 1843 sheet music of "My Early Home" was Edwin Scrantom's sentimental recollection of his family's original log cabin at Lot No. 1 at the Four Corners. Nostalgia could only be enjoyed once such primitive days were past.

Fig. 31. (*top*) 1838 view of the Eagle Tavern in Rochester on Lot No. 1 at the Four Corners. This brick building represented a substantial upgrade over the frame tavern it replaced. The Bank of Monroe rented quarters within this prominent corner building.

Fig. 32. (*bottom*) Photograph of the Eagle Hotel, c. 1850, at the Four Corners. During the wave of second-generation refinements, the Eagle was upgraded in multiple ways. Along with the elevated appellation of "hotel," the building was architecturally upgraded with classical porticoes. Interior refinements separated the sexes into their own saloons and parlors. This building was replaced in the 1860s by the Powers Block.

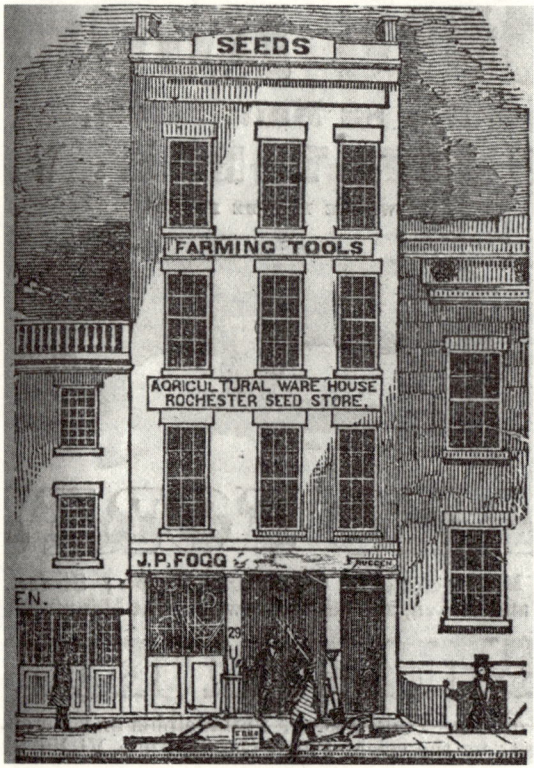

Fig. 33. (*top*) 1850s engraving of the male milieu of a Philadelphia tavern. Men lounge on the rails and chairs of the front veranda, while inside they drink at the bar and read the news.

Fig. 34. (*bottom*) 1850s advertisement for a Buffalo Street business, Rochester. In terms of its architecture and social sorting, the store is typical of those constructed during the second quarter of the nineteenth century. Cellar establishments, including barbers, oyster sellers, and wholesale departments, typically catered to men; note the top-hatted man emerging from the cellar business next door.

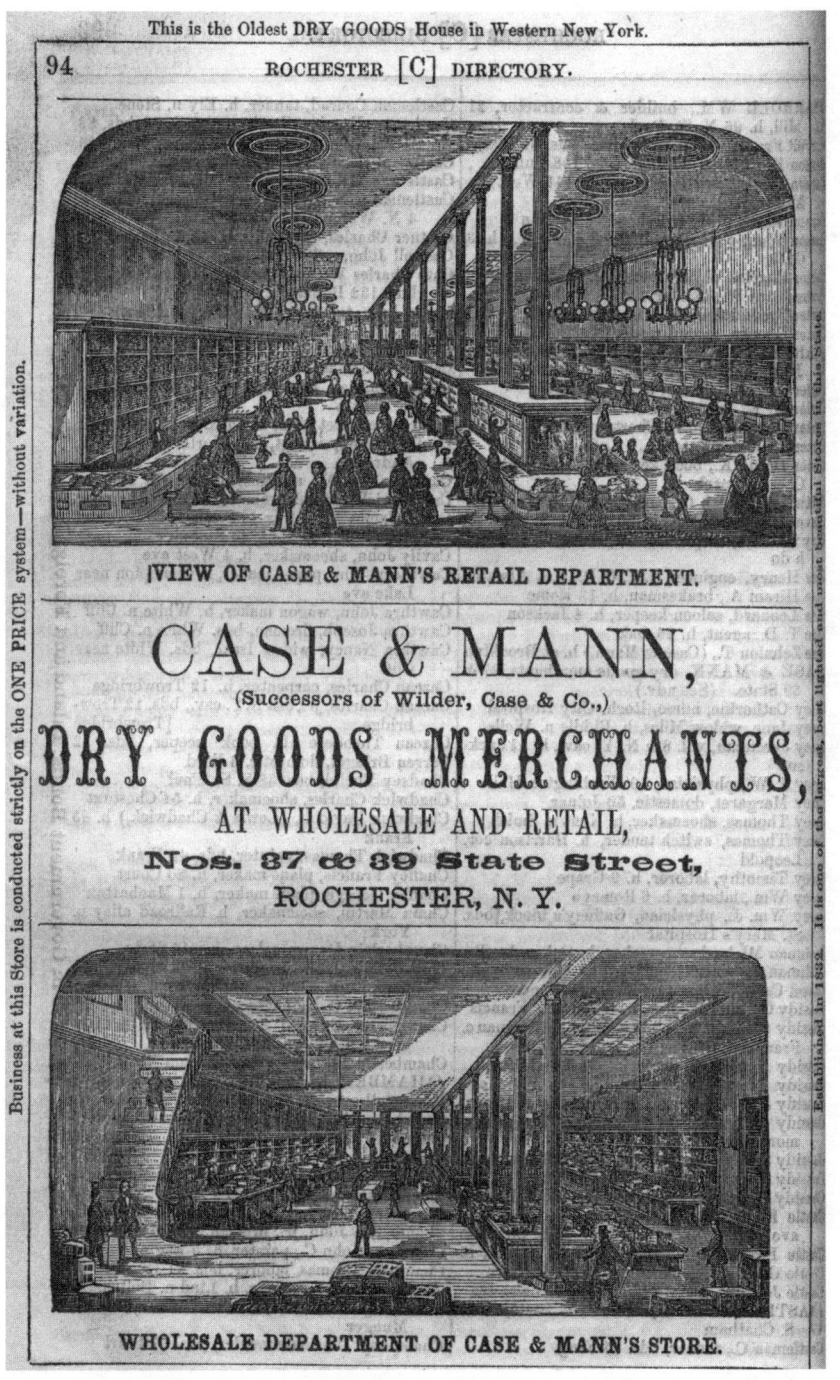

Fig. 35. 1860s advertisement for Case & Mann dry goods store on State Street, Rochester. The wholesale store is in the basement and serves a male clientele of sales agents and jobbers. The retail store at street level has a higher degree of polish and is oriented toward the female shopper.

Fig. 36. (*top*) 1837 engraving of bookseller shop. The glass display window makes this a visually permeable shop and, not coincidentally, a shop that women frequent.

Fig. 37. (*bottom*) 1869 typesetter's icon for a grocery. At many stores the price of goods and customer's treatment depended on the clerk's appraisal of the shopper's gender, class, and race.

Fig. 38. (*top*) 1860s advertisement for Ellis' Music and Piano Room. The plush, parlorlike setting borrowed from the language of domesticity, suggesting a suitable environment for females. Depicted here is the 1863 Rochester reception for Tom Thumb and his bride, Lavinia Warren.

Fig. 39. (*bottom*) 1837 engraving of a millinery shop. Shops owned by and catering to women tended to seek secluded quarters in the upper stories of business rows.

Fig. 40. (*top*) 1837 engraving of a lithography shop. The upper-story workshop is devoid of architectural embellishment.

Fig. 41. (*bottom*) 1865 advertisement for Bryant, Stratton & Folsom's commercial college, Troy, New York. Mercantile colleges training young men commonly rented quarters in the upper stories of buildings in the commercial district. Bryant, Stratton & Company was a large chain, with mercantile colleges throughout cities in the East and Midwest.

Fig. 42. Yates Clothing Store (1865), Syracuse. Just north of Clinton Square across from the Empire Block's dry goods row, this business block evinces the vertical integration of the new ready-to-wear clothing industry, with production and sales under one roof.

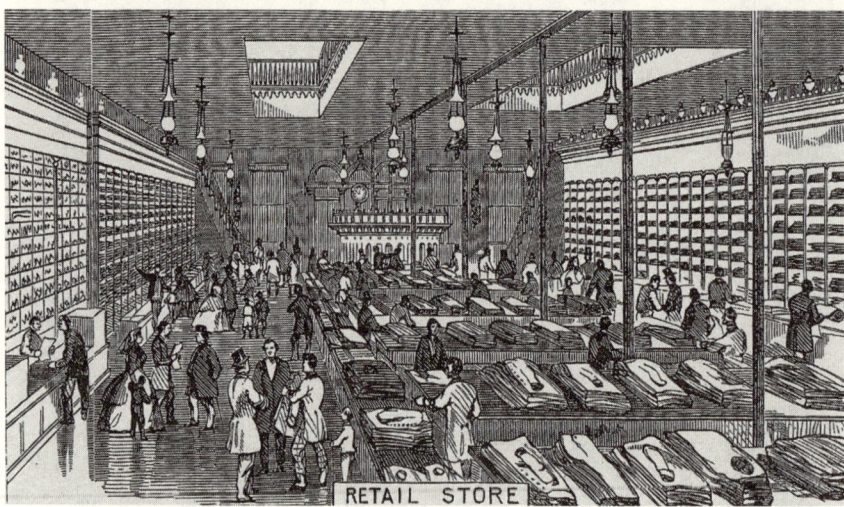

Fig. 43. The first-floor retail and second-floor general wholesale room in the Yates Clothing Store. Carrying men's clothing only, both the retail and wholesale sales floors were oriented toward a male clientele and escorted females.

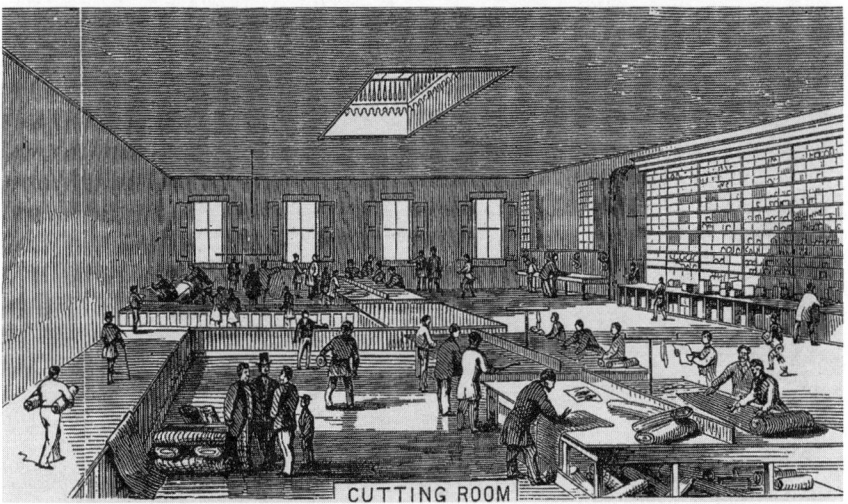

Fig. 44. The third-floor cutting room and fifth-floor manufacturing room in the Yates Clothing Store. Gender-segregated production spaces were removed to the upper stories. Female operatives were sequestered on the very top floor in cramped and supervised conditions.

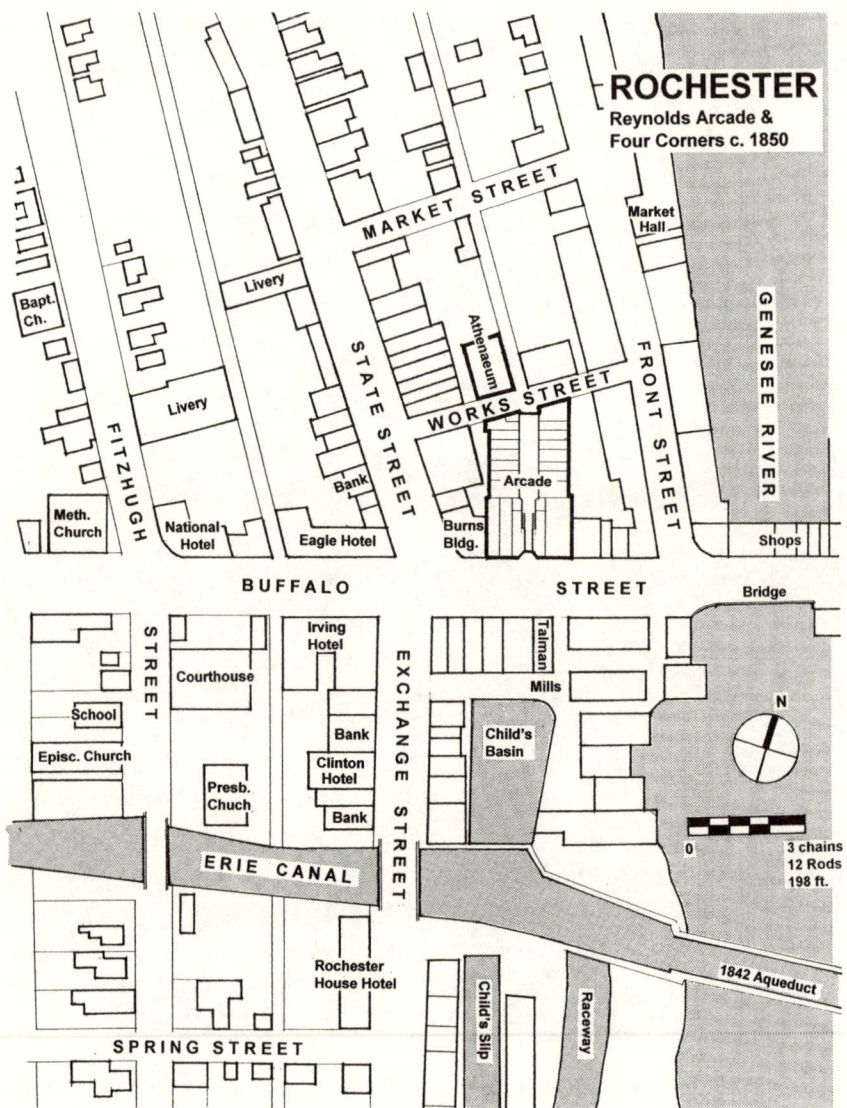

Fig. 45. Map of Rochester's Four Corners. By midcentury Reynolds Arcade had been extended all the way from Buffalo Street to Works Street. Reynolds Athenaeum further exploited the real estate potential of the interior of the block while also gentrifying its social milieu.

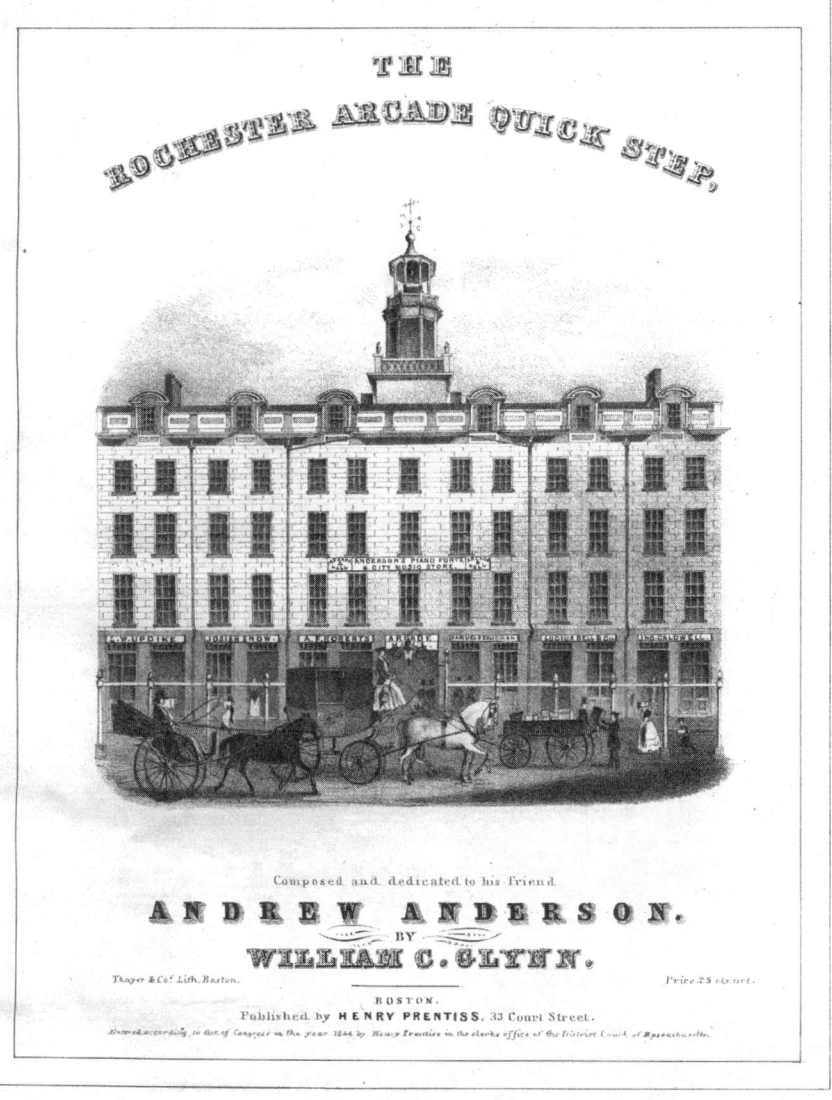

Fig. 46. 1844 "The Rochester Arcade Quick Step" immortalized the building in sheet music. A place-making edifice, the Arcade "stamped our individuality when we were hardly expected to have individuality," recalled one resident.

Fig. 47. 1851 view of Reynolds Arcade interior showed the polite tradition of the promenade, with retail shops below and professional offices above.

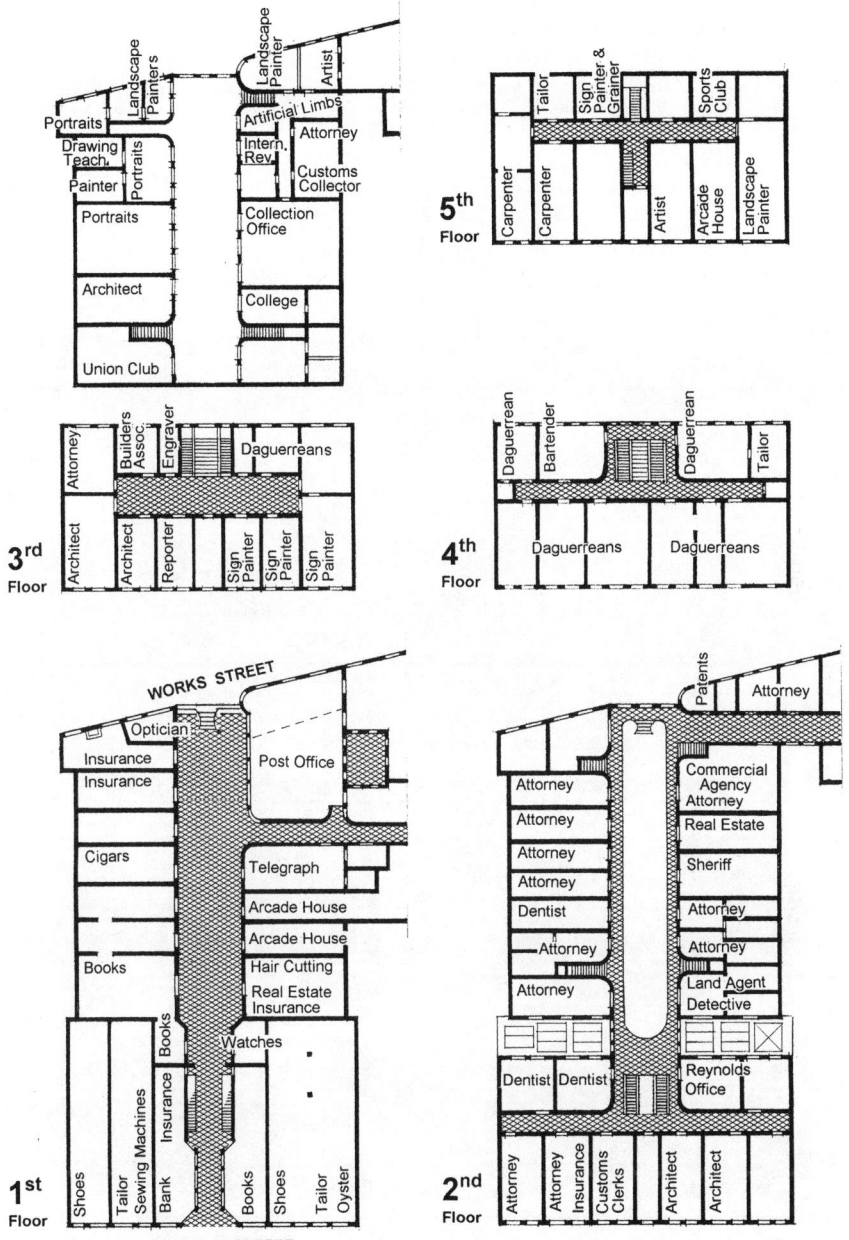

Fig. 48. Plan of the Arcade in 1863. By the 1860s the Reynolds had lost tenants to the dry goods row on State Street, but the Arcade remained an important location for professional offices. The Arcade was also renowned for its unique tradition of housing an artists' colony in the upper floors.

Fig. 49. 1861 view of the Athenaeum, also known as the Corinthian Hall after its top-floor assembly room. Urban, architectural, and social gentrification went hand in hand with this 1849 construction project behind the Arcade.

Fig. 50. (*top*) 1851 view of the Athenaeum's elegant Corinthian Hall interior outfitted for a public banquet during the Agricultural Society Fair. Classical detailing inspired its name as well as its elegant reputation.

Fig. 51. (*bottom*) 1880 view of Works Street, known before the Athenaeum as the seedy Bugle Alley. As reworked, the rear facade of Reynolds Arcade, including the post office wing, faced the Athenaeum and presented a genteel commercial streetscape.

Fig. 52. (*top*) 1856 view of Syracuse showing the railroad crossing at South Salina Street. The Greek Revival train shed straddles Washington Street.

Fig. 53. (*bottom*) 1878 photograph of Wolf and Salina streets in Salina. The Salina neighborhood where the McCarthy's originally ran their dry goods store failed to keep up with the economic or architectural progress of the Syracuse commercial district.

Fig. 54. Nineteenth-century photograph of South Salina Street between Washington and Fayette streets. McCarthy's dry goods store moved to this corner location in 1850; McCarthy established a separate wholesale store near Jefferson Park in 1876.

Fig. 55. 1852 bird's-eye lithograph of Syracuse, looking northwest. Picnickers look over the new city beneath them. Clinton Square is dominated by the white L-shaped block of the Empire House hotel and business block. The slanting parallel lines of the canal and railroad approach from the right. Rows of saltworks fade in the distance toward Onondaga Lake. Fayette Square in the foreground became an elite residential neighborhood. Exaggerated church steeples suggest a moral citizenry.

Fig. 56. Detail, 1874 bird's-eye lithograph of Syracuse. The Armory (3) in the oval of Jefferson Park anchors the reclaimed millpond and saltworks. Train tracks run along Washington Street; a feeder line cuts diagonally through the lower left-hand corner. The McCarthy's new warehouse will shortly be built at the intersection of Washington and Clinton streets (76). The courthouse (2), Empire House hotel (77), and Syracuse House hotel (75) edge Clinton Square. The weighlock is opposite the new city hall (4) just visible in the upper left-hand corner. Salt boiling sheds line the Oswego Canal at the upper left, and the movable roofs of the solar evaporation sheds are barely visible in the lower left-hand corner.

Fig. 57. Detail, 1853 bird's-eye lithograph of Rochester, looking east. Buffalo Street remains the axis for the civic and commercial districts. The cupola of the Bethel Church rises near the junction of canal and street. The blocky 1851 courthouse stands tall with its domed cupola. The tower of the Episcopal church is superimposed upon the Presbyterian steeple. The thicker congregation of people and wagons in the street marks the intersection of the Four Corners. The Arcade and Athenaeum are barely distinguishable among the dense cluster of businesses at Buffalo and State streets.

ONE

Vernacular Urbanism and the Mercantile Network of New Cities

Vernacular urbanism: Both a type of urban settlement and a method of urban analysis. This approach focuses on the ordinary city rather than the extraordinary metropolis. It emphasizes nonacademic design rather than professional planning and social implications rather than formal analysis. Vernacular urbanism interprets a cityscape as a cultural construct whose streets and buildings result from both the physical construction and social production of urban space.

Sorting: The intentional process of ordering activities, buildings, and people within urban space in response to a dominant culture's notions of good urban form. In the nineteenth century, a pervasive mercantile ethos resulted in new cities whose public spaces of commercial, industrial, and civic activity were geographically, functionally, architecturally, and socially sorted to enhance a self-fulfilling image of economic vitality and bourgeois gentility.

During the early nineteenth century westward expansion was not only pushing the frontiers of the nation through the Midwest but was also redoubling on itself as the undeveloped portions of the eastern colonial states became more intensively settled. Accelerated by the opening of the Erie Canal in 1825, white settlement of the New York interior had actually begun shortly following the American Revolution with the clarification of state boundaries, the extinguishing of Native American land rights, the government distribution of bounty lands for military service, and the entrepreneurial reselling of vast tracts of land. The state's demographics were radically shifting. Seventy-five percent of New York's population clustered along the exclamation point of New York City and the Hudson River Valley in 1785, but by 1820 the ratios had reversed, and three-quarters of the population now lived to the north and west of Albany.[1]

The roving journalist William Leete Stone toured the curious new settlements of upstate New York in 1820 (Fig. 1). Stone observed firsthand that urban beginnings were messy and questionable. Traveling along the first completed stretch of the Erie Canal, his packet boat emerged from the dusky wilderness into Syracuse, a clearing hacked out of the swampy forest. As the last rays of daylight bleakly illuminated the settlement, Stone spotted a scattering of shabby wooden houses constructed amid the stumps of recently felled trees. Marooned for the night, he lodged at a miserable tavern frequented by local salt boilers whose "wild visages, beards thick and long, and matted hair" presented "a group of about as rough-looking specimens of humanity ever seen." It snowed that dreary October night, and the next day he found "the morning aspect of the country even more dismal than the evening before."[2] Syracuse's chief booster, Joshua Forman—the lawyer, land agent, and partial proprietor who had recently commissioned a new urban plat—pressed the chagrined traveler for his impressions. "Mr. Forman," castigated Stone, "Do you call this a village? It would make an owl weep to fly over it!" "Never mind," replied the optimistic Forman, "you will live to see it a city yet!"[3]

The built environment could undermine an urban identity, but it could also promote one. Despite necessarily primitive beginnings, mythologized in genesis stories such as these, the new nineteenth-century cities along New York's frontier quickly displayed the principle of sorting, which guided their layout and permeated subsequent developments. By agreeing with his dubious visitor, the booster Joshua Forman showed that he too was culturally literate about how to read a city and, by extension, was knowledgeable about what his new city needed in order to be legible. Although it was imperceptible to Stone, rough-edged Syracuse was already taking sorted shape. An inn at the crossroads and a sawmill by the stream staked out two separate nodes for commerce and industry. But the new city lacked the critical mass of either buildings or settlers to make its urban armature or aspirations apparent.

A mere nine years later Stone would return and declare that upstart Syracuse had been transformed: "Another enchanted city, I exclaimed, as I glanced upward and around upon splendid hotels and rows of massive buildings in all directions—crowded too, with people, all full of life and activity! . . . For if noble ranges of buildings, two or three large and tasteful churches, busy wharves and streets, and all the life and animation of a large commercial place will constitute a city, then, most assuredly Syracuse may be called by that name."[4] Being without a municipal charter until 1848, and thus not a city in the legal sense, Syracuse had nonetheless attained the status of a city in Stone's eyes. It was not simply the architecture but also its arrangement into discernible nodes. Stone had observed three sorted districts: a bustling commercial landscape of hotels and business rows, a transport landscape of wharves and canal-side warehouses, and a civic landscape of churches and county buildings. Although closely spaced within the confines of a small city boasting perhaps two thousand residents, these three functions were legibly discrete and thus had pushed Syracuse across the urban threshold.[5]

The fact that these new nineteenth-century cities were ordered and sorted—functionally, architecturally, and even socially—forces a reconsideration of the urban typologies and processes that have been commonly used to categorize nineteenth-century urbanization. Building upon the work of Sam Bass Warner, Jr., historians have traditionally identified two American urban archetypes bracketing the nineteenth century. First came the small-scaled, socially and functionally mixed cityscape—typically called the "walking city"—which hallmarked the compact, pedestrian-scaled city. By the end of the century the walking city had evolved into the large metropolitan cityscape of distinct commercial, industrial, and residential districts, including "streetcar suburbs" that were enabled by technological and transportation improvements.[6]

The walking city model has undergone continuous reappraisal. Warner originally acknowledged that there was some geographic sorting by function within the walking city: "for all its conglomeration of little buildings, the early big city was not a disorganized hodgepodge."[7] But much of the focus on sorting has been on the definition of residential districts, particularly the rise of bourgeois residential neighborhoods, where class consciousness became an impetus for sorting.[8] Early commercial buildings were often combination shop houses, and this mix of residence and work has blurred the presence of functionally discrete districts, leading to the erroneous interpretation that, as long as work and home were combined in one premise or even in the same neighborhood, there was little, if any, functional and spatial differentiation.[9] This assumption is incorrect. Even when people lived where they worked, there could still be occupational segmentation, such as commercial districts, milling districts, or warehousing districts. The shop, not the house, was the primary attribute that identified the district. The assumption that residential use contaminates or lessens the resolution of a commercial or industrial district is a twentieth-century, post-zoning construct, not a nineteenth-century tenet. Historical studies that have focused on social sorting and housing variety have, nonetheless, valuably advanced the idea that the compact, preindustrial city may have been more sorted than first thought.

The walking city paradigm nonetheless turns out to be an inapt description of the pedestrian-scaled, highly sorted cities that were created de novo during the nineteenth century. The walking city model is predicated upon the examples of mature, East Coast, major metropolises with colonial underpinnings. The nineteenth-century inland city sorted into commercial, industrial, and sometimes civic districts arose out of a different set of conditions. Their very youth set the new cities apart. Their newness was part of the challenge to city founder and settler alike, but it was also a rich opportunity. After visiting the new city of Rochester, one traveler concluded that "progress, there, cannot fail to be rapid, when, in addition to manifest physical advantages, still imperfectly developed, the blood of the whole city is *new*, untied and untrammelled by old notions, or hampered by forefather prejudices."[10] The marvel of seemingly overnight urbanization in the interior led many travelers, including the English woman Fanny Trollope, to exclaim that "Rochester is one of the most famous of the cities built on the Jack and Bean-stalk principle."[11] As speculators staked out new towns across a wooded landscape, these new settlements on America's eastern frontier were windows into a new era of city building and a new opportunity to rethink the role and design of cities.

Built, paradoxically, within a nineteenth-century frontier of a seventeenth-century state, New York State's sorted cities were constructed physically on a white man's tabula rasa but construed conceptually as part of an established political economy that led back to Albany and New York City. A generation before places such as Syracuse or Rochester were laid out, the state and speculators were preparing inland New York for economic development. New turnpikes stretched across the southern, middle, and northern tiers of the state. A surveyors' net of squared counties and townships checkered the landscape. A smattering of settlements were already planted, and investors marked promising sites for urban and economic development. As the social historian F. C. Wallace has summarized, "by the time the common 'pioneer' himself arrived, the land had been explored, bought from the Indians, garrisoned, pacified, partially surveyed, mapped, and picked over by land companies, public officials, and private speculators seeking the most probable localities for quick profit."[12] Being neither isolated outposts in the far west nor fully integrated cogs in the Atlantic economy, the new frontier cities of New York were commercial ventures consciously designed to create profits. The state legislators who financed transportation improvements, the autonomous city founders who platted the new settlements, and the independent settlers who built them envisioned the new cities as links in a mercantilist market chain being forged from one end of the state to the other.

The new cities Syracuse and Rochester were part of a common vernacular landscape of independent urban speculations in the New York interior. In size and type both cities represent a more common urban category than the metropolitan stars that dominate urban historiography.[13] By 1820 there were only five cities in the United States with populations between 25,000 and 125,000. There were, however, another 56 cities numbering from 2,500 to 24,999, including Rochester, with 2,063 residents, but excluding Syracuse, which had probably fewer than 50 settlers. A decade later Rochester had risen to 9,207 residents, and Syracuse boasted 2,565.[14] Indeed, nineteenth-century urbanization can be characterized as the efflorescence of new and small cities. Using the conservative measure of 2,500 inhabitants, an urban threshold used in midcentury censuses, the number of cities rose dramatically from 33 in 1800 to 2,128 by 1900. Only a small percentage of this increase was the result of colonial villages growing into cities. Most of the growth was caused by the creation of completely new settlements. Of the 124 largest ranking cities in 1890, only 19 had exhibited a measurable and independent existence in the 1790 census.[15] Rochester and Syracuse were two of those newly emergent cities.

The founding of Rochester began in 1803, when three business partners from Hagerstown, Maryland, joined forces to purchase from the Pulteney Estate a failed mill site on one hundred acres of land on the west side of the Upper Falls of the Genesee River, just seven miles from its discharge into Lake Ontario. By the time Nathaniel Rochester, William Fitzhugh, and Charles Carroll finally gained clear title in 1811, they had expanded their plans from reviving the mill site to planning an entire city within the nearly square boundaries of the so-called 100–Acre Tract. Having moved to New York State in 1810, Nathaniel Rochester served as project overseer, platting the city and earning the honor of naming the great venture that he actively promoted (Fig. 2).[16]

The genesis of Syracuse began in 1804 when the state sold 250 acres of land to Abraham Walton of Utica in order to finance the construction of turnpikes that would help develop the freight potential of its Salt Springs Reservation as well as open the marshy area for settlement. Located a few miles south of Salina, which had been laid out in 1798 as the administrative center for the state's salt reservation, the irregular shape of the Walton Tract reflected creative surveying designed to include as much dry land as possible. In fulfillment of the state's sales stipulation, Walton had an inn and mill constructed, but land speculation rather than actual settlement was the order of the day. The slowness of the settlement to take hold is indicated in its frequent name changes in vernacular usage. As a nod to the very first innkeeper, the settlement was known as "Bogardus Corners" until 1809, then grandly as "Milan" until 1812, then in the shadow of Salina as "South Salina" until 1817, then in reference to the new innkeeper as "Cossitts Corner" until 1817, and then classically as "Corinth" until 1820. Needing a name unduplicated elsewhere in the state in order to get a post office established, the local investor John Wilkinson tapped into the popularity of classical place names and proposed the name Syracuse in reference to the ancient Italian city that similarly had salt springs and a freshwater lake. Development of the boggy settlement lagged until the state announced the routing of the incipient Erie Canal through the village. In 1819, under the direction of the Onondaga Hollow resident and land promoter Joshua Forman, the old tract was resurveyed, platted into a core of urban lots and a periphery of farm lots. Matters looked promising, and Forman moved to the rejuvenated experiment of Syracuse that year (Fig. 4).[17]

Vernacular Urbanism and Spatial Culture

Syracuse and Rochester, the new cities that are the focus of this study, developed in a way best explained within the context of what may be called "vernacular urbanism." This term borrows from the subject and method of vernacular architecture studies, with its emphasis on the ordinary over the extraordinary, nonacademic design over professional or tutored design, and social over formal analysis. Syracuse and Rochester were, if not quite ordinary in their success, certainly part of a large wave of new city speculations during the early nineteenth century. Their focus on commercial and industrial opportunities put the cities squarely within the territorial and mercantile expansion of the United States. Although the opening of the Erie Canal in 1825 secured their fortunes, even as flourishing canal towns they were part of a familiar city building phenomena of seemingly instant sensations. Following his Erie Canal tour, the author Nathaniel Hawthorne surmised, "surely, the water of this canal must be the most fertilizing of all fluids; for it causes towns . . . to spring up."[18]

In addition to identifying a common urban type, the label of vernacular urbanism implies a process of design which is not limited to the purview of trained specialists. These new cities are certainly part of the history of urban planning, but they were not designed by professional planners. Rather, they were created by entrepreneurs working with and as surveyors to lay out an urban plan. Nathaniel Rochester and Enos Stone worked side by side in laying out Rochester's plan for his city. Joshua Forman similarly directed his brother Owen Forman and associate John Wilkinson in the 1819 laying out of Syracuse. These city founders were tutored neither in the art of design nor in the economics of development but were de facto planners by dint of their experience on the land and their plans for the future. Just as vernacular architecture can imply non-architect design, vernacular urbanism can imply nonplanners.

Vernacular urbanism also connotes a method of historical analysis which investigates the social constructions behind spatial practices, or what may be called "spatial culture." Spatial culture can be understood as a type of equation. One factor is the physical city: the relatively static, formal arrangement of buildings and lots and street patterns. The other factor is the human interaction within the urban container: the fluidity of use and meaning inscribed on the built environment by the ways in which people used (or were inhibited from using) that space functionally, socially, and culturally. Together the physical and the social combined to create the sum that is spatial culture, the aggregate organizational principles that governed urban space.[19]

This beginning definition of spatial culture is not quite complete, as it addresses the processes that shaped the city but not the objectives behind it. In working toward a spatialized approach to urban anthropology, Setha Low uses the mutually complementary perspectives of the "social production" and the "social construction" of space. The social production of space, Low argues, includes "social, economic, ideological, and technical [factors] whose intended goal is the physical creation of the material setting."[20] In contrast, the social construction of space is "reserved for the phenomenological and symbolic experience of space as mediated by social processes such as exchange, conflict, and control."[21] These perspectives add the important element of intent behind spatial practices by recognizing that there are no absolute or neutral spatial practices since all actions emerge from an ideological base, whether consciously acknowledged or not. Spatial culture acknowledges physical space as an active agent in the shaping of notoriously ethereal social relations. It equally acknowledges the ways that human behavior can challenge or reinforce the meaning of physical space within the city. Examining the spatial culture of the new nineteenth-century city injects human agency and cultural conditions into understanding the design of cities and buildings.

The spatial culture of the new city reflected the depths to which the commercialization of the landscape had permeated the social construction of urban society in antebellum America. Analyzing the spatial culture of the new cities provides a method to conceptualize the forces at work which shaped the sorted cityscape. New York's chain of towns, loosely gridded plans, divided districts, architectural refinements, and even standards of public behavior within public space were forged by culturally conditioned choices. Despite being founded and settled by quite autonomous and unrelated individuals, the similarity between the functionally, architecturally, and socially sorted spatial cultures of Rochester and Syracuse suggests a common culture of city building common to the eastern frontier. More specifically, the spatial hierarchies inherent in the functionally, architecturally, and socially sorted cityscape suggest the power of the merchant and professional classes and the presence of dominating commercial motives in the discourse and practice of city building at the time. The sorted cityscape highlighted a presumption of economic diversity, market efficiencies, extra-local connections, and urban maturity. The sorted cityscape was both style and substance. Its distinct commercial, industrial, and eventually civic districts articulated the merchant-led economic and cultural colonization of the hinterland.

The tripartite sorted cityscape was not atomized by its separate districts but, rather, was connected and defined through them. The conceptual term *cityscape* connotes a perceptible set of relationships which constituted the whole of the city in its own time.[22] Contrasting the ideas of "land" versus "landscape," the cultural geographer J. B. Jackson concluded that all "scapes" are collections of related things organized as parts of a greater system with a sense of purpose to and interrelatedness among all the parts.[23] A scape, therefore, is an arranged environment, be it a landscape, townscape, or cityscape. But the suffix *scape* implies not only the calculation to arrange the parts but also a recognition of the pattern. As a concept, a cityscape is not simply the sum total of the city's parts. It is the mental organization of classified parts and hence a subjective creation on the part of both the founders who conceived it and the observers who perceived it. Both the intentional arrangement and the reflective acknowledgment of the schema sets a cityscape apart from a city or a landscape apart from land. The sorted cityscape was a cultural artifact meant to be understood by its diffuse audience.

Although a landscape is no longer limited to its original meaning as a type of painting, there is nonetheless a pictorial element in the creation of any scape. A scape is a picture in the mind's eye. The organizing system within a cityscape includes a degree of aesthetic consciousness. It is something that is perceived by those who shape, view, or experience the totality of the place. Both to construct and to comprehend a cityscape require reflection and a shared pattern of language. Since a cityscape is space arranged in a way that is meant to be understood by visitors and residents, the clearest cityscapes will therefore use the most common, legible, and easy to understand constructions and arrangements. Nonetheless, a cityscape is always a subjective interpretation of urban space, architectural form, and social meaning. This is the paradox of cityscapes, since cities are notoriously untidy and incomprehensible in their totality: the intentions behind cityscapes are varied, the interrelatedness of the parts are not always readily apparent, and the people who are creating and perceiving the cityscape are often themselves anonymous or invisible. Yet cityscapes are created, organized artifices.

The new city's sorted cityscape was an assemblage of pieces of urban space knitted into a coherent whole. Sorting was a system of classifying space into individual pieces and then recognizing the connections among those spaces.[24] The sorted city was not an accident. Anything constructed by humans is the result of decision making, and environments are no exception. The presence, the location, and the appearances of the districts were products of a series of deliberate decisions made from the time of platting through the actual construction of buildings and use of city space. While the rise of commercial and manufacturing districts in particular is often dismissed as a natural evolution, space does not sort itself out; people do the sorting.[25] A landscape, as John Stilgoe convincingly asserted, "happens not by chance but by contrivance, by premeditation, by design."[26] So, too, a cityscape, including one that was sorted.

The Commercial Culture of Improvement and the Rise of the Merchant's City

New nineteenth-century cities led by a productive merchant and manufacturer elite turned the antiurbanist's arguments on their head. Despite Thomas Jefferson's famed warning that the great cities were "pestilential to the morals, the health and the liberties of man," New York's new cities were more commonly heralded by politicians and businessmen as champions of economic progress which in turn instilled social and cultural improvement.[27] The antiurban rhetoric of the corrupting influence of the city was being challenged by two new moral imperatives of particular appeal to city builders: the twin virtues of creating economic productivity and carrying civilization. The commingling of these two principles eventually found its voice in the expansionist doctrine of Manifest Destiny but very quickly influenced the substance and style of discussions on improving the state's hinterland.[28] Contemporaries championed cities as vital contributors in developing the agricultural countryside. As commercial entrepôts, cities made agriculture itself profitable while also increasing the market reach of merchants and manufacturers. As social centers, they touched the isolated farmsteads and encouraged a more refined urban culture in the settled areas. Cities were vehicles of improvement.

During the wave of territorial expansion following the American Revolution, cities played a major role in the settlement of the Midwest and the infilling of the eastern frontier.[29] In his 1827 anniversary address to the New York Historical Society, Joseph Blount extolled the urban transformation of both the state and the nation: "Instead of several distinct communities, thinly scattered through thirteen provinces along the sea-coast, we find a dense and united population pouring into the interior, accompanied by the arts of civilization, and the refinements of social and cultivated communities. Educated and intelligent man is taking the place of the savage, and is fast advancing to the borders of the Pacific ocean, making the wilderness to smile like a garden, and sowing towns and villages as it were broadcast through the country."[30] A generation later the public was still amazed at the ongoing transformation of New York. A passage written in 1848 by the Virginian J. C. Myers following his journey through upstate New York illustrated a similar appreciation for urbanization as a vehicle for a moralized economic improvement: "I viewed with surprise the numerous handsome and flourishing cities, towns, and villages which I passed through since leaving the city of Schenectady. . . . My surprise was still increased, when informed that they were more numerous from Rome westward, . . . Here then I had another opportunity of reading in bold characters, the noble enterprise and industry of the northern man."[31] Myers was silent regarding the physical form of these new urban centers, but his word choices of *enterprise* and *industry* doubly resonate. These heroic platitudes, synonymous with diligence and effort, captured the idea of hard work justifiably rewarded by profit and progress. The descriptors *enterprise* and *industry* also began to limn the very commercial and industrial activities that steered urban development. As templates for development, these two economic activities would be inscribed into the plats of new cities.

The idea of improvement permeated public discussions on settlement in general and urbanization in particular. While touring Rochester in 1827, the Englishman Basil Hall had opportunity to ruminate on the term:

> It may be proper to remark, that about this period I began to learn that in America the word improvement, which, in England, means making things better, signifies in that country, an augmentation in the number of houses and people, and, above all, in the amount of acres of cleared land. It is laid down by the Americans as an admitted maxim, to doubt the solidity of which never enters any man's head for an instant, that a rapid increase of population is, to all intents and purposes, tantamount to an increase of national greatness and power, as well as an increase of individual happiness and prosperity.[32]

Improvement was a multivalent term that carried connotations of personal progress in intellectual, spiritual, or social matters as well as connotations of economic or productive progress through some sort of value-added process of work. Noah Webster's nineteenth-century American dictionary revealed the nuances of the word.[33] The first definition, "Advancement in moral worth, learning, wisdom, skill, or other excellence," captured social, spiritual, and intellectual values. The second, "Melioration; a making or growing better, or more valuable; as the improvement of barren or exhausted land; the improvement of the roads," touched on the physical. Another definition explored the idea of best and highest use: "Use or employment to beneficial purposes; a turning to good account." Webster's *improvement* blended work, productivity, progress, and righteousness.

During the nineteenth century this complex notion of improvement permeated the Anglo-American capitalist culture of land use and became an articulated legal value that was protected under the American judicial system. Overriding other variables in evaluating property rights cases, the courts consistently gave preference to the best and highest use of a property, to economically productive activities over non- or less-productive ones. In so doing, the courts effectively subsidized the nation's economic growth by privileging profitable utility (or improvement) over other activities or values.[34]

The civilizing and moralizing aspect of "best and highest use" infiltrated the public discourse on settling the frontier with the assumption that productive and profitable landscapes were more righteous than were unimproved landscapes and that the greater the impact, the greater the good.

Commercialized agriculture was better than subsistence agriculture. Commercialized towns centrifugally feeding a mercantile network were superior to country towns serving a centripetal market. This line of reasoning fed into a stream of urban justifications that argued cities were correctives to a dereliction of economic and cultural opportunity and thus were agents of an improved civilization. City building was acquiring a moral tinge. In fact, Webster tapped directly into urbanization's link to physical and social progress when he expounded on his fourth definition of *improvement* with the statement "I look upon your city as the best place of improvement."[35]

Joshua Forman's oration at the 1825 Syracuse celebration of the Erie Canal captured the spirit of the age, linking the issues of frontier settlement, transportation, civilization, and mercantilist revenue all under the rubric of *improvement*:

> To what extent this course of improvement may be carried, it is impossible for any mere man to conjecture; but no reasonable man can doubt, that it will continue its progress, until our wide and fertile territory shall be filled with a more dense, intelligent, and happy people. . . . It has long been the subject of fearful apprehension to the patriots of the Atlantic States that the remote interior situation of our country (for want of proper stimuli to industry and a free intercourse, with the rest of the world) would be filled with a semi-barbarous population uncongenial with their Atlantic neighbors, but the introduction of steamboats . . . and canals . . . promise the wide spread regions of the west all the blessings of a sea-bord district. But while we contemplate the advantages of this work, as a source of revenue to the State, and of wealth and comfort to our citizens, let us never forget the means by which it has been accomplished.[36]

The rise of the market economy in eighteenth- and nineteenth-century America contributed to this instrumental use of cities to promote regional development. Commercial and manufacturing towns were envisioned as links in a wider mercantile chain of improvement. The market, and the related topic of transportation, dominated much of the public discussions on the settling of central and western New York, influencing the vision of cities and, ultimately, their physical form. Frontier landholders framed the issues of settlement around the production and consumption of resources and goods, setting the commercial tone for developments. They called for the state to invest in internal improvements that would specifically enhance the commercialization of agriculture and with it the commercial and urban development of central and western New York.[37] As early as 1791, the New York governor was exhorting the state legislature to approve internal transportation bills, arguing that the rapid increase in frontier settlements "must yield extensive resources and a profitable commerce," provided links could be established.[38]

Politicians, developers, and settlers agreed that transportation was the key to unlocking the interior's potential for commercially conditioned physical and cultural improvement. The first calls were for roads; then came the outlandish proposals for a statewide canal linking Buffalo on Lake Erie to Albany on the Hudson River and thence to the New York City harbor and the Atlantic world. Governor DeWitt Clinton continually stumped for public support for the Erie Canal, painting a picture of a veritable floating market: "boats loaded with flour, pork, beef, pot and pearl ashes, flaxseed, wheat, barley, corn, hemp, wool, flax, iron, lead, copper, salt, gypsum, coal, tar, peltry, ginseng, bees-wax, cheese, butter, lard, staves, lumber."[39] Clinton's vision of bountiful productivity explicitly carried biblical conviction as well. Quoting the Old Testament prophet Isaiah, Clinton predicted that the canal would transform New York State, "the wilderness and the solitary place will become glad, and the desert will rejoice and blossom as the rose."[40]

The rhetoric of commercial opportunity pulsed through inland urbanization. Cities and transportation went hand in hand in maximizing the economic potential of the whole state. An 1816 canal proposal noted, "new markets will be opened by increasing population, enlarging old and erecting new towns, augmenting individual and aggregate wealth, and extending foreign commerce."[41] A western New Yorker pointed out that linking "the eastern markets and the interior . . . would secure to our own people the full benefit of their industry, their traffic and commerce, and build up our own towns and seaports."[42] These newly built-up towns would be the hinges between production and consumption, stimulating and serving the demand for goods and services within their regions and ultimately linking western and eastern markets.[43] When Henry O'Reilly wrote his four-hundred-page *Sketches of Rochester* in 1838, he included a chapter on "Progress of Improvement" which gave credit to the highways, canals, and railroads.[44] A toast at the opening of the Erie Canal in 1825 wittily pointed out America's productive pragmatism: "To America, cutting canals while Europe is cutting heads."[45]

City founders paid particular attention to the existence or possibility of transportation improvements that would link their settlement to the nation's market chain. Although both Rochester and Syracuse were created before the Erie Canal was planned, the road-rich cities subsequently capitalized on their place along the "artificial river."[46] At the Rochester opening of the Erie Canal, the city builder Nathaniel Rochester offered his toast: "the greatest publick [sic] work in America, if not in the world. A principal link in the chain that binds the Union of the States."[47] The Syracuse booster Joshua Forman similarly toasted: "our village is the offspring of the canal and, with the county, must partake largely of its blessings."[48]

Promoters envisioned roads, canals, and later railroads as tools for maximizing potential across the state and as solutions to the mercantilist desire to keep profits within the state. The agricultural quickening of western New York with Genesee wheat and the industrialization of the state salt reservation in central New York was under way by the late eighteenth century, but the lack of convenient markets diverted profits and stymied production. In 1807 the surveyor Benjamin DeWitt devised an organic analogy to illustrate the market relationships of eastern and western New York and the necessity of reliable transport in maintaining the state's "body economic": "Let us consider the city of New-York as the centre of commerce, or the heart of the State, Hudson's river as the main artery, the turnpike roads leading from it as so many great branches extending to the extremities, from which diverge the innumerable small ramifications or common roads into the whole body and substance; these again send off the capillary branches, or private roads, to all the individual farms, which may be considered as the secretory organs, generating the produce and wealth of the state."[49] New roads or canals linking the western and central portions of the state with Albany (and thence down the Hudson River to New York City) would redirect the flow of profits that had previously found easier outlets via Lake Ontario to British Canada or via the Susquehanna River to Philadelphia.[50]

Mercantilist improvement was not the only objective behind urbanization. Social and cultural improvement also factored into the discussions of a commercialized landscape threaded with roads and canals. The white, propertied, merchant class developed their argument that commercial cities were positive agents of social change, first by eliminating Native Americans and then by disciplining slovenly frontiersmen. J. C. Myers, who had been so struck by New York's urban chain, heroicized the efforts and products of his fellow white, male, Americans: "The noble enterprise of the white man has so changed the aspect of this region, that upon every hand attractive beauty meets the eye."[51] His physical evidence was not a landscape but a cityscape. "Here now far and wide the aboriginal forest has lost its charms of savage wildness, by the beauties of cities, towns, and villages and the intrusion of railroads and canals."[52] The eradication of the Native American landscape, conceived by whites as a landscape of underutilized and lost opportunity, and its replacement with the merchant and manufacturer's cityscape became a tautological defense for urbanization.[53]

In lobbying for transportation investments, a Buffalo landowner and congressman warned against a lopsided social and economic landscape that had "merchants, manufacturers, and agriculturists" in the eastern half of the state but only farmers in the west.[54] Boosters argued that urbanization enlightened the scattered white settlers who had grown slothful through poverty and isolation on their farms. In 1810 an Erie Canal promoter linked moral salvation and economic opportunity: "The object of a wise legislature is to promote industry and virtue in the state, but we know that people who live far from market, and cannot sell their produce, naturally become indolent and vicious. Having little to do, they do less.... There are people in the western country, settled on a bountiful soil, who do not raise a bushel of grain except what is eaten by the family, or what is made into whiskey, for the purpose of drowning thought, and destroying soul and body."[55] In an 1832 lecture entitled "The Moral and Other Indirect Influences of RailRoads" the speaker argued that "society alone is a rich source of education and improvement ... [and] society [is] better in cities than in towns, in towns than in villages, and in villages than in country places."[56] The rural countryside could no longer be counted on as the locus of moral or cultural supremacy, concluded the editor of *Harper's New Monthly Magazine* in 1855. "It is plain that the great things in history have not been done in the country," he argued; instead, it was the city where one celebrated "the triumphs of literature, of art, and benevolence."[57]

A commercial, capitalist urbanism was seen as a civilizing, democratizing, and moralizing force. The physical and temporal overlap of New York's "burned over district" during the Second Great Awakening of Protestant revivalism and the early-nineteenth-century commercial infilling of interior further blended the two perspectives into a moralized economy of development. Markets and morality combined in the new virtue of capitalism. Indeed, city building became imbued with a certain religious and social righteousness. Recalling the first twenty years of Rochester's history, one of the first residents proudly wrote: "The place grows rapidly about us— and instead of the wilderness that frowned here in 1813, a *city* now is here teeming with people and with business, and spires are seen, as it were pointing the wayward and thoughtless to a better & more permanent city, 'it made with hands eternal in the heavens.'"[58] Cities, according to this argument, were not Jefferson's feared sores on the body politic but were fruitful contributors to national strength. When *Putnam's Monthly Magazine* declared in 1855, "the great phenomenon of the Age is the growth of great cities," it was with pride, not apprehension.[59]

The similarity of choices made by the city founders and settlers shows how deeply the culture of commercial development had infiltrated ideas about the role and form of cities. The sites for Rochester and Syracuse offered their own natural advantages, but their city founders envisioned them as more than processing centers for the region's raw resources. Rochester abutted the great milling power of the Genesee River Falls, a great asset in a wheat-growing region, but its founders chose a grander scheme than a simple mill site. Syracuse was created out of land sold by the state in order to finance road construction, yet the Syracuse promoters envisioned a more ambitiously urban future than a simple way station. Conceived within an emerging commercialization of the landscape—commercialized agriculture, commercialized processing of raw resources, and commercialized settlements to export western and import eastern products—city builders platted these new cities to tap into the growing market economy.[60] They laid out cities designed for local, regional, and distant commerce and industry and sought county seat status to boost their political and economic centrality within their markets.

Given the rhetoric of commercial settlements contributing to the economic and moral improvement of the interior, it is little wonder that the white male merchant entered as the figurehead in public discussions and public culture. The nineteenth-century business publisher Freeman Hunt lauded the urban merchant as the agent of commercialism who not only civilized but also democratized the landscape. Preaching to the choir, he pointed out that commerce—unlike the medical, legal, or ministerial professions—was itself a democratic profession that could be practiced by all without specialized training. Moreover, the profession provided a service to humanity by creating wealth. That wealth could then be reinvested in the community. "It is evident," Hunt concluded, "how much we owe to Commerce, and how greatly we depend upon our merchants, for our means both of social progress and religious effort."[61]

The self-interested reciprocity between mercantile city planting and the merchant class who planted them is perhaps not surprising but does merit deeper investigation about the ways in which these specifically bourgeois class interests shaped the physical and social construction of the new city.[62] The new ideology of commercial improvement created both a physical and a discursive space for a new, commercial elite. Merchant interests played a large part in constructing the economic colonization of New York; as a result, the history and culture of the new cities were deeply rooted in entrepreneurship, and the cities they bullied into existence bore their stamp economically, politically, physically, architecturally, and socially. Social and political rewards went to those entrepreneurs who risked their economic capital in making the urban venture a success, either as city founders, merchant millers, or merchant shopkeepers. Economic investment translated into social and political opportunity, prerogative, and privilege.

Entitlement born of economic and political power further translated into social and cultural hegemony as well as spatial privilege within the city. "It is not stating the truth too strongly to say that America is proud of her merchants," wrote Hunt, "in fact it is another name for gentleman among us."[63] The historian David Scobey uses the term *bourgeois urbanism* to capture the importance of class in shaping the city of New York, specifically that class of "propertied, professional, and genteel" men who held the wealth, political power, and cultural authority that permitted them to conflate nation building, city building, and class formation in one grand gesture.[64] That cultural authority extended to the hinterland as well. In his study of the urbanization of rural New England the historian Richard Brown has contended that a business class—composed largely of merchants, millers, lawyers, printers, and artisans—"exercised a disproportionate role in bringing urban culture to the countryside."[65] More specifically, he argued that they integrated the Second Great Awakening's emphasis on individual spiritual regeneration with a secular component that included active citizenship, self-development, and cultural engagement with the wider world—all of which endowed them greater confidence and even authority in local matters.[66] Paul Johnson's study of Rochester's class relations richly demonstrated the depth to which the emerging merchant class, whose outlook blended a businesslike, urban, bourgeois, and religious tinge, dominated the city.[67] In 1842, after visiting America, "that vast counting-house that lies beyond the Atlantic," Charles Dickens concluded that it seemed as though "everybody is a merchant" sharing a "national love of trade."[68] While not a universal sentiment, this love of trade was held by the dominant class of white, male, businessmen whose economic and cultural status was in ascendancy and who had led the mercantilist city-building efforts since the turn of the century.

The rhetoric of commercial urbanization which surrounded the infilling of New York's frontier made only the barest of allusions about the form these new cities would take. The attention was on what the cities should do, rather than what they should look like or the manner in which they should be organized. But the dialogues about transportation, agriculture, commerce, manufacturing, and social improvements pointed some city builders in the direction of a cityscape functionally sorted by the type of improvement.

The commercialized idea of improvement was inscribed into, indeed abetted by, the urban, architectural, and social fabric of the new cities. The commercial elites who largely dominated the rhetoric and financed the actual building schemes of inland settlement imprinted their notions of good urban form on the landscape. Good urban form began as cityscapes sorted by function, a sorting that maximized the land's economic utility and advertised its economic opportunity. Under the leadership of the merchants and millers, good urban form shouldered additional responsibilities of being a managed social landscape as well. The sorted city was the product of a commercially oriented, merchant-led initiative to improve the interior. Urban morphology, building type and style, and the social use of public space were all manipulated to create a merchants' city that suited their complex urban agendas. A sorted cityscape provided a powerful template and rhetorical tool in demonstrating a new city's balanced and integrated economy as quickly as possible. The commerce of retail and wholesale trading, the industry of milling or shipping, and even the social improvement of churchgoing or civil governance could all be given physical form, as evinced in William Leete Stone's recognition that "noble ranges of buildings, two or three large and tasteful churches, busy wharves and streets, and all the life and animation of a large commercial place" constituted a city."[69] An Albany newspaper echoed these sentiments in its 1824 account of the new interior cities, whose commercial, industrial, and civic buildings were beacons of economic and cultural progress: "Your eye is regaled with the most beautiful scenery. At one time you pass through an old settlement, and again you witness nature just submitting to the strong hand of cultivation. You observe the progress of art from the log hut of a squatter, the illegal settler; the comfortable farm house; the village just bustling into existence, and flourishing town with its gilded spires, bustling streets, and active industry."[70] The cityscape was proof of improvement.

TWO

Planning the Sorted City
Commercial, Industrial, and Civic Districts

"The rage for erecting villages is a perfect mania," marveled DeWitt Clinton in 1810 over the rash of urban speculations in the New York interior.[1] Indeed, Syracuse and Rochester were hardly lonely outposts. The growing commercial cities of Buffalo, seventy-three miles west of Rochester, and Utica, sixty-one miles east of Syracuse, had a decade head start and higher population figures to prove it.[2] Thirty miles south, smaller county seats such as Auburn, Geneva, Canandaigua, and Batavia were strung along the Ontario and Genesee Turnpike.

Even more pressing, Syracuse and Rochester faced rivals in their immediate vicinity. Syracuse was overshadowed by nearby Salina, which had been created by the state in 1798 as the local administrative center for the state salt reservation, and by Onondaga Hollow, which had become the county seat in 1794. Nearby Manlius, Pompey, and Cicero were laid out following the sale of the Military Tract lands. The single-industry salt-boiling towns of Geddes and Liverpool also shouldered Onondaga Lake. Rochester was one of six new settlements that speculators had staked out along the final seven-mile drop of the Genesee River. Charlotte, refounded in 1805 on the site of a failed 1792 settlement, was at the mouth of the river on Lake Ontario. Hanford's Landing, refounded in 1807 on the site of the abandoned 1796 Fall Town at King's Landing, was located about three miles farther upriver, below the fifty-seven-foot Lower Falls. Carthage, platted in 1817, sat above the Lower Falls. Frankfort was laid out in 1807 at the ninety-six-foot Middle or Main Falls. Rochester, platted by Nathaniel Rochester in 1811, was laid out at the twenty-foot Upper Falls a half-mile farther upriver. East Rochester, first occupied in 1811 then platted by Elisha Johnson as a village in 1817, faced off directly across the river.[3] Nathaniel Rochester also worried about other new towns being laid out farther upriver, "which makes it the more necessary that we should do the same in order to turn the attention of people to the most eligible spot."[4]

In order to attract settlers and investors, Syracuse and Rochester needed to be seen as places that offered superior economic opportunities, services, and conveniences. To become commercial cities with a complex internal economy of mercantile and manufacturing activities, they needed to develop, cement, and extend commercial relationships with both hinterland and eastern markets. All this, of course, would take time, but time was an ill-afforded luxury. In order to attract the human and financial capital needed to build a city and an economy, the new settlements needed to be able to display a promising urban agenda as quickly and persuasively as possible.[5]

The city plans of Rochester and Syracuse overwhelmingly subordinated civic interests to commercial ones.[6] Responding to a cultural climate that stressed economic improvement and commercialized landscapes, the city founders imprinted a mercantile cast onto their plats (Figs. 2–5). The commercial intentions were evident not only in what was included in the plats but also in what was excluded. Neither Syracuse nor Rochester radiated around an honorific civic square that could give literal or figurative prominence to the idea of civic community.

Instead, the ninety-nine-foot-wide commercial thoroughfares that crisscrossed both Rochester's Four Corners and Syracuse's Clinton and Hanover squares were the most prominent form of open, public space in the cities. The open spaces of rectilinear Clinton Square and the wedge of Hanover Square, which on paper resembled formally designed and community-oriented civic squares, were actually intersections of crossroads and the Erie Canal and functioned essentially as commercial widenings given over to markets and wharves (Fig. 22). Rochester's Four Corners, the intersection of Buffalo, Carroll, and Mill streets, was occasionally called a square, even though each corner was built upon and the only thing open was the street (Fig. 32).[7] This Four Corners "square" was the symbolic heart of the city. Nathaniel Rochester's decision to label the large northwest lot "No. 1" was itself a statement that this was the point of origin for the whole city. Nathaniel Rochester did reserve a block of lots beyond the Four Corners for a county courthouse, but that cluster of lots was neither a full square nor a central focus. Similarly, in Syracuse the juncture of the canal with Salina and Genesee streets (also known as "the Corners") became the central point for establishing the cardinal coordinates for streets and building address numbers. In both Syracuse and Rochester these squares and corners were for the movement of goods and people, and not a place for a civic monument or institution, civic assembly, or recreational enjoyment for the community.

These wide thoroughfares—edged by towpaths, sidewalks, and shops—became the sites and symbols of public life in the new commercial city.

This crabbing of civic, public space stood in sharp contrast to many colonial cities as well as to the choices made in other contemporary new cities. Legislated into existence during the early nineteenth century, new Middle Tennessee county seats pragmatically focused on their centripetal courthouse square, which demonstrated the very raison d'être for the birth of the town. The speculator's new town in the Mid- and Far West was similarly designed around a public square, this time awaiting some institutional privilege. As some new cities evidenced, civic squares did not necessarily require public buildings as anchors; they could be designed as communal open spaces. The avowedly commercial gridiron of 1830s Chicago included a designated public square intended not for a public edifice but, rather, as open public space "forever to remain vacant of building."[8] Even the agricultural common of colonial New England survived a nineteenth-century commercial transformation of the landscape, becoming an urban green linking commercial and civic activity.[9] The eighteenth-century "Pennsylvania town" type with its notched-out central diamond presented another option that created an open square articulated by commercial and civic buildings. The Pennsylvania town model would have been personally known to the founders of Rochester. Nathaniel Rochester met his partners William Fitzhugh and Charles Carroll while living in Hagerstown, Maryland, a county seat that featured a classic Pennsylvania town diamond at its center. Some of the new cities in New York State experimented with prominent open central squares set within precisely gridded plans, including Brown Square in Frankfort, the tract immediately north of Rochester, and Washington Square in Salina, the three-mile distant rival to Syracuse (Figs. 3 and 5). Both rival settlements enjoyed a superior economic advantage—Frankfort had the largest falls, and Salina had the monopoly as the state's salt administration center—but their plans literally turned their backs on the industry at the borders and focused attention inward on the nonproductive, civic central square.

As the city builders of Frankfort and Salina would quickly learn, civic space and industrial potential meant little if a rival had better plans for integrating a new settlement into the commercialized landscape. In fact, topography and ineffective politicking conspired against the settlements. Both Frankfort and Salina failed to win key positions along either the new state roads or the Erie Canal. Despite the fact that Salina's salt springs and Frankfort's thundering falls gave unparalleled natural advantages to their sites, the civic-squared towns grew weakly and were annexed into Syracuse and Rochester, respectively. Certainly, Frankfort and Salina did not fail because of their squares, but the presence of the squares can be seen as their founder's distraction from a dedicated commercial vision. Conversely, the omission of civic space, the emphasis on the commercial street, and the successful pursuit and integration of transportation improvements worked together to maximize the legibility and viability of Rochester and Syracuse as blossoming mercantile cities.

The prominence of major thoroughfares played into the commercial posturing of the new cities. City founders understood that transportation was crucial to the success of a settlement and relinquished valuable real estate acreage for transportation paths. The ninety-nine-foot-wide main streets did more than facilitate commerce; they made a statement about the importance of commerce. And these bold statements could be influential. Arriving in Syracuse in search of opportunity, Marcus Hand was seduced by the broad intersection of Salina and Genesee streets. The young man decided that their "airy appearance" distinguished the settlement from other places with less urban potential, and he decided to stay and seek his fortune.[10]

The actual construction of the sorted city would depend on the consensus of the settlers who built up the city, but for now the founders could control the sorting through platting. Their plans manipulated four elements that blended style with substance and aspiration with actuality. First, the planners prioritized integrating regional transportation routes that would expand their settlement's industrial and commercial possibilities. Second, they configured their plats into variations on a grid, a type whose straight lines and expansive potential signified a city. Third, the articulated weave of the cross streets provided a clear armature on which planners and settlers could apportion the commercial, industrial, and civic districts that were the functional proofs of urban complexity. Fourth, the planners manipulated street and lot dimensions, prices, and availability to encourage compactness, the critical mass needed for a visibly and functionally credible urban endeavor. By laying out compact and sorted cities whose plats embraced strategic transportation routes, the city founders advertised both the intention and the means by which their ventures might grow into vital urban centers.

Transportation and Regional Networks

Neither Nathaniel Rochester's 1811 plat for Rochester nor Joshua Forman's 1819 plat for Syracuse were abstract designs veneered on a blank landscape (Figs. 2 and 4). Their geometric plans balanced the surveyor's convenience of the grid with the economic necessity of integrating transportation routes.[11] On paper the plats appear self-contained, bounded by a defined perimeter with the emphasis on the deep interior anchored by a major intersection. The arteries did not even have to be in actual existence; the solid promise of an incipient road or canal was promotion enough in the earliest days. These intersections of actual or predicted road and canal lines formed the nucleus of each settlement and the heart of the commercial district. The apparently centripetal focus of the crossings belied the greater centrifugal truth behind the platting of these new cities. The roads and canals that seem so vital to the center of town were only important because they extended beyond the platted city, connecting it to a larger world. Roads and canal dominated the design of the sorted city, not just inside the plat but as prerequisites for platting. These new cities could not exist without regard to their regional context. Indeed, transportation was critical to the viability of commerce, industry, and even county governance. The crossroads defined the urban core, but its extension beyond the city limits defined the city's place within a larger regional network of cities and hinterlands. In a sense these were cities designed from the outside in.

Syracuse was born of transportation and had roads and canals before it ever had much to move on them. Initiated as the product of a turnpike financing scheme at the turn of the century, it was bolstered a generation later by the Erie Canal. During the 1790s a road connected the Salina saltworks to the north and the village of Onondaga to the south. The state carved out a 250–acre tract for sale in 1804, specifically to raise money for an extension of the Genesee and Ontario Turnpike. As instructed by the state, the purchaser Abraham Walton built a sawmill on the Genesee Turnpike next to Onondaga Creek and contracted Henry Bogardus to open an inn at the crossroads of the road to Salina and the new Genesee Turnpike (hence, the settlement's early name of Bogardus Corners). Thus began the process of building industrial and commercial districts. The crossroads settlement remained poorly developed even after being passed down through several speculators. Unlike Rochester, whose location by the falls placed it in a prime position to be an active industrial site, Syracuse did not have the same refining or processing advantages. Its sawmill was basically a support system for area salt producers, who used its lumber for the boiling sheds, evaporation vats, and storage barrels as well as for other new construction. Bogardus Inn was no great site either, being the place where William Leete Stone made his sorry stay among the local salt boilers.

After intensive politicking by central New Yorkers, fortune smiled in 1817, when the state legislature finally approved and funded the unprecedented canal proposal and routed its course through both Syracuse and Rochester.[12] Anticipating a burst of interest in this transportation node, Joshua Forman, who had been elected to the state legislature on a "canal ticket" and who owned property in the lagging crossroads community, hired Owen Forman and John Wilkinson to plat a new city in preparation of intense speculative interest. The importance of Salina Street, Genesee Street, and the crossing Erie Canal is manifested in the 1819 plat, which centered around their convergence at Clinton Square primarily and the smaller Hanover Square secondarily. A superimposed grid of secondary streets knit the veering lines of the arteries together into a reticulated plat. The feeder line of the Oswego Canal extended the reach of Syracuse to Salina and thence northward to Oswego and Lake Ontario, further tying the city into wider markets.

The Erie Canal was a phenomenon that carried local, state, and ultimately national implications for its extension of the mercantile network. Weighlocks augmented the importance of their host cities as shipping points and toll collectors. The state constructed only seven weigh stations, including one at Syracuse and Rochester, along the course of the Erie Canal. A weigh station brought commercial advantages to its host city by creating local jobs at the weighlock as well as commercial opportunities for those catering to the canallers stopped for assessment.

Rochester, too, was transport dependent, but as a parallel instead of primary criteria for development. The waterpower of the Genesee River's Upper Falls had already enticed investors with the hopes of a mill site, but there was not serious consideration of an urban settlement until the Maryland trio contracted in 1803 to purchase the "100–Acre Tract." Rumors had circulated as early as 1805 that a canal might cross the Genesee River above the Upper Falls, but a more realistic opportunity at the time was the routing of the new Buffalo-bound state road that skirted the shores of Lake Ontario. By 1809 it had stopped just a few miles east of the Genesee River.[13] Not yet having clear legal title to the contracted 100–acre purchase, Nathaniel Rochester nonetheless successfully petitioned the legislature for the bridgehead, arguing that the shallow ford immediately above his Upper

PLANNING THE SORTED CITY 25

Falls provided one of the simpler crossings. He then used the proposed state road, appropriately named Buffalo Street, as the east-west axis when laying out the new settlement. When the Erie Canal was cut through the city during the early 1820s, it nearly paralleled Buffalo Street, confirming the east-west economic flows of the city. Carroll Street, named in honor of one of Rochester's partners, crossed Buffalo Street (and later the canal) at an angle paralleling the course of the unnavigable Genesee River. The northward course of Carroll Street led to the harbor at Lake Ontario and, by extension, wider U.S. and Canadian markets. The new city may have been small, but with these transportation options its market reach was vast.

The construction of the Buffalo state road was plagued by intrigue, proving that much was at stake in the town-building sweepstakes. Town promoters did what they could to beg, borrow, or steal a transportation advantage. Fearing the loss of western trade, businessmen from Canandaigua on the southerly Genesee and Ontario Turnpike argued that the northern road should also dip south toward them and denigrated Rochester as "a God-forsaken place! inhabited by muskrats, visited only by straggling trapper through which neither man nor beast could gallop without fear of starvation or fever and ague."[14] After the legislature approved the crossing, rival town leaders concluded that insults were ineffective and tried new tactics. In 1811, as the state road approached Rochester and headed for the assigned river crossing, an unexplained diversion occurred. On a chance site visit from his new home in Dansville, about forty miles away, Nathaniel Rochester discovered that the road inching toward his paper city suddenly took an abrupt angle and headed north toward the rival new towns of Charlotte and Carthage. Working through the state authorities, Nathaniel Rochester forced the road to return to its legally chartered course. The crook in the east side's Main Street (including the bend of old Main Street, or today's East Avenue) is the legacy of this correction (Fig. 3).[15]

Roads and canals were obviously powerful agents in abetting commerce and manufacturing. They were also imperative in gaining and retaining county seat status. County seats that were ill situated for the convenience of the wider county population could be stripped of their status, and a more central town selected as the new county seat. Good lines of transportation, therefore, could either protect a county seat or help a challenger win that status.[16]

Transportation was important not only to business but to the business of settlement as well. The new roads and canals carried both goods and people. Potential settlers were scouting the region for opportunities, and travelers undertaking an Erie Canal or Niagara Falls journey were similarly paying attention to the sights along the way. Located on state thoroughfares, Rochester and Syracuse became de facto tourist and scout sites. As travel writing became a popular literary genre, these disseminated accounts began to frame the reputation of a place.[17] When stopping in Rochester in the early 1830s, the emigrant scout John Fowler, loaded down with travel accounts, recorded that the city "fully answered every representation I had heard of it."[18] For new cities trying to garner some attention, it was imperative to be on an important thoroughfare. State roads, canals, and railroads were place-making devices that promoted, rather than simply responded, to development.[19]

Grids

With the regional network marked out and the local crossroads delineated, the planners could turn their attention to filling in the rest of their plats in a manner suited to a commercial city. Any plan was actually a promising sign of urban aspiration because a premeditated design separated the man-made environment from unimproved, and hence devalued, nature. The plat alone could be a compelling document of urban aspiration. Spafford's 1824 *New York State Gazetteer* included Lodi, one and a half miles east of Syracuse, as if it were a real place, even though "it is as yet but a village on paper."[20] Platting inspired confidence on the part of purchasers as well. So much so that they were willing to pay "city" prices for as yet unimproved land. Nathaniel Rochester charged thirty to fifty dollars for a quarter-acre lot in his paper city, at a time when nearby agricultural land was selling for only five to fifteen dollars per single acre.[21]

The 100–Acre Tract rose in value with each speculation but soared after urban platting. In 1802 Ebenezer Allan sold his failed mill site to the Pultney Company for $7.50 per acre, who sold it the following year to the Maryland partners of Rochester, Fitzhugh, and Carroll for $17.50 per acre. The partners pledged to stay the course, despite some initial setbacks. In 1811 Charles Carroll urged "that nothing should induce us to divest ourselves of the fee in any part of that property . . . you can fully attain my ideas on its rapidly increasing value and scarcely any sum would weigh with me one moment to divide or part with it." Carroll was adamant: "Hold fast this property—take my advice, never part with it. Hold it—it is an estate for any man."[22] In 1812, when the partners commenced the sale of individual lots in Rochester, the price per acre calculated out to $120 to $200.

Syracuse real estate escalated with the coming of the Erie Canal. The price per acre had remained steady in the crossroads village of mill and tavern between 1804 and 1816, the year when Joshua Forman and his partners paid $4,500 for three-quarters of the original Walton Tract. Unable to complete the purchase of the full tract before word of the Erie Canal got out, the partners had to raise another $3,000 in order to purchase the remaining one-quarter of the tract. Unable to capitalize quickly enough on their debt, they were forced to sell off their holdings. By 1824 an Albany-based consortium known as the Syracuse Company had paid $30,000 for the bulk of those lots.[23]

Far from being perfectly orthogonal, the skewed grids of Syracuse and Rochester balanced the topographic realities of accommodating the roads and canal while adopting the practical advantages and symbolic potency of grid planning. The plats that were exciting such attention in Rochester and Syracuse were variations on the grid, a plan whose straight streets expressed an urbane and urban agenda. When Daniel Drake of Cincinnati wrote in 1815 that "curved lines represent the country, straight lines the city," he was presenting rectilinear planning as the appropriate standard for cities.[24] A generation later Ralph Waldo Emerson similarly invoked linear geometry to compare the city and the country, writing that "the City delights the Understanding. It is made up of finites: short, sharp, mathematical lines, all calculable."[25] Contemporaries prized grid plans as a form that expressed not just urbanity but a particularly modern, rational urbanity. Before being overcome by the monotony of its repetition across the country, the Englishman Francis Baily praised the "perfect regularity" of Philadelphia and Baltimore, noting that grids were "by far the best way of laying out a city. All the modern-built towns in America are on this principle."[26] The straight line was attractively unnatural. Its artifice showed the hand of humans in altering the spontaneous meander into a purposeful plan. In contrast to the rationality of the grid, colonial cities such as Albany garnered criticism for their narrow and crooked streets.[27]

The grid was an especially pragmatic choice for a commercial city. Grids were not only convenient for surveying and selling lots; they were also easily expandable, a quality of particular interest to city builders looking to grow their settlements. In 1811, the same year that Nathaniel Rochester laid out his city, the commissioners of New York unfurled a massive grid across Manhattan as a means of encouraging orderly and efficient platting, selling, and building of lots in the commercial metropolis. The wide avenues anticipated a northward flow of traffic, while smaller cross streets linked the two waterfronts. A universal grid removed the future entanglements of political and vested interests in particular tracts of land on the island. Its rhetorical flourish of republican equality also represented a classicizing taste for formal symmetry and balance. Even New York City, however, was permitted a deviation from the grid in the interest of transportation—the colonial Boston Post Road, better known as Broadway, continued its angled run through Manhattan. The history of the 1830 platting of gridded Chicago was particularly reminiscent of Syracuse, since it too was born of the promise of a canal and the state applied its land sales to finance the transportation project. Chicago was also laid out in a virtually limitless grid.[28]

Sorting

The founders of Rochester and Syracuse believed that a successful commercial city demanded more than a grid and major transportation routes. They also believed a new city should be arranged to allow for economically and spatially distinct districts. In the hands of the city planners the grid was not a neutral system that treated all space equally; rather, it was a tool for defining difference. Indeed, city planners built in preconditions that would promote functional sorting of commercial, industrial or shipping and, in the case of Rochester, civic districts within a compact area. The loosely gridded plats of Syracuse and Rochester provided a clear physical armature that ordered urban space, and hence activities and experiences. The emphasis on commercial and industrial districts in particular made a strong statement about the ideals behind the settlement.

Older merchant cities in England and America already featured functional sorting. London's medieval craft and merchant guilds had established occupationally defined streetscapes, and the later capitalist development of the city and its colonial offspring refined that sorting into what the geographer Martyn Bowden calls "the mercantile triangle."[29] The sorted pattern of the mercantile triangle can be envisioned as stacked bands of commercial activity tapering inland from the waterfront. The wide base of the triangle was the waterfront, where the shipping and commission activities of merchants and wholesalers operated. Moving inland (or farther up the triangle) were financial services with offices, insurance companies, and exchanges. Farther inland was the retail shopping area, followed by the entertainment district, and, at the triangle's apex, the civic district of legal or cultural institutions. Implicit in this mercantile model is a merchant nucleus that principally excluded artisan craft production, manufacturing, and residential use. The fluidity of the American seaports, such as Philadelphia, Boston, Baltimore, with their rapid turnover of population and buildings, made it even easier socially and spatially to construct a mercantile triangle in a way that their prototype of entrenched London could not. Even in these American metropolitan centers, however, the triangle could be flattened to reflect thin resources, with several functions compressed within a single band.[30]

Smaller cities compressed the sorting even more. In 1837, when the business publisher Freeman Hunt described River Street in the new city of Troy, New York, as "the Pearl, the Front, the Water street, and the Broadway of Troy," he was alluding to the bands of wholesale, financial, and retail activity which, usually given fuller geographic expression in the metropolis, had been condensed into one important street in the small city next to the Hudson River and Erie Canal terminus.[31] Yet the consolidation of these functions still points to a sorting impulse, as River Street excluded other functions such as civic institutions or housing. Moreover, Troy's River Street, which paralleled the Hudson River, was functionally divided down the middle into two bands of commercial subdistricts. Enterprises that did not need the river's wharves—such as inns and local retail stores—tended to locate on the inland, land side of River Street. In contrast, the water side of the street was taken up by stores and warehouses that could utilize the river frontage. These "waterside" buildings were in turn subdivided into street-side and river-side businesses.[32]

The plans of other new turn-of-the-century cities sorted functions morphologically as well. The complex plats of Washington, D.C. (1792), Buffalo, New York (1804), and Detroit, Michigan (1807) gave great diagrammatic clarity to functional sorting. These plats were extraordinary examples of grand manner planning, with radiating baroque diagonals converging on plazas of civic importance set apart from commercial areas. The planners of lesser towns were also experimenting with complex ways to sort functions. A 1794 proposal for Esperanza, New York, eclectically combined different grids, a circular square, and converging radial avenues to articulate separate functional zones of a courthouse square, a church square, a market street, and wharves.[33] These complex plats allocated specific functions into their own discrete, articulated space and are indicative of the value placed by city planners on sorting in the new nineteenth-century city. Notably, all of these new ventures emphasized a tripartite cityscape of commercial, industrial, and civic functions.

Few city builders, however, had the ability or desire to invest in such idealized and artful extravagances of grand manner planning, particularly when similar sorted results could be had with the expedient grid. The commissioners who devised the gridded extension of New York City in 1811 explicitly rejected those "supposed improvements, by circles, ovals, and stars" which embellished a plan but at the cost of "convenience and utility."[34] The open spaces and plan adjustments dedicated to artistry, grandeur, or even the promotion of healthfulness came at too high of a cost. Certainly, fancy plans were more difficult to lay out, as surveying manuals made clear.[35] The alternative costs of open plazas or public gardens were also too expensive, because they consumed valuable real estate that could have been sold for profit. When such space-consuming or inefficient elegancies were actually implemented, as in the geometric extravaganzas of Circleville, Ohio, and Jeffersonville, Indiana, they often could not be sustained in the face of pressures for land and convenience, and their complex plans reverted to simpler grids.[36]

In both Syracuse and Rochester the founders created a sorted cityscape within the grid by directing particular functions to particular places through direct and indirect suggestion. These city builders implemented their own variations on the sorted mercantile triangle that privileged the merchant's economic landscape. Befitting inland cities, the roads and canals served as the waterfront but in a way that permitted development along both "fronts," modifying the triangular pattern the seaboard merchant's cityscape. Retailing and wholesaling activities were clustered along the cross streets. Shipping and commission activities, augmented by milling, formed one district fronting the canal or falls. Banks straddled the two economic landscapes, being located around Hanover Square in Syracuse and on Mill (later renamed Exchange) Street in Rochester. And county courthouses, when finally gained, were located at the outer edge of the commercial district, in the far corner of Clinton Square in Syracuse and one block away from the Four Corners in Rochester.

Rochester's grid was the most straightforward and the most emphatic example of diagrammatic clarity (Fig. 2). This was to be a commercial city, built for profit and attractive to business interests. Nathaniel Rochester focused on the four blocks radiating from the Four Corners as the urban nucleus. He sorted commerce into the northwest and northeast blocks, milling into the southeast block, and civic activity into the southwest block. His 1811 plat showed his conviction that a successful city would be achieved by implementing a compact and sorted cityscape that displayed the triad of commerce, industry, and civic functions. By designating specific activities for the four blocks, Nathaniel Rochester was doing more than just assigning functions; he was laying the groundwork for a larger and more complex sorting process that would ultimately lead to architectural and social distinctions within the cityscape.

In the case of the new cities early industrial practices included the activities of processing raw materials, warehousing the products, and shipping goods for export and import. The Rochester milling district was the most straightforward to lay out, and Nathaniel Rochester even explicitly named this southeast quadrant the "mill lots." The Upper Falls and its dilapidated millrace had been the initial impetus behind the partnership's purchase of the 100–Acre Tract, and they continued to dictate the location of the new mill seats. Although the logic of the site was obvious, partner Charles Carroll still felt compelled to remind Nathaniel Rochester not to "sell any ground that can in any possible manner or shape injure or interfere with any sites or situations for water works."[37] The partnership also retained control over the mill sites by renting rather than selling them, thus ensuring how the lots would be used. Nathaniel Rochester further reiterated the mill district's purpose by naming the key street that lined its outer edge "Mill Street." Denoting the absence of milling in the northern blocks, the street name changed as it crossed Buffalo Street to "Carroll Street" in honor of the founder Charles Carroll. Nathaniel Rochester's careful land policies, bolstered by a terminology of *mill lots* and *Mill Street*, made it clear what was, and what was not, part of the mill district. Clearly, this mill district was not to be diluted with other functions.

Rochester also specifically designated the blocks by the mill race for "building warehouses on the river," a note that alluded to expanded shipping and commission opportunities that should result from his city's competitive transportation situation.[38] These new cities were intended not as agricultural villages with closed or localized economies but, rather, as outward-looking commercial towns pursuing opportunities along the mercantile chain. Thus, the milling landscape was composed of not only mills but also warehousing and freight facilities that extended the mills' reach. Reliable and efficient shipping was requisite for effective industry; therefore, roads and canals were important components to the industrial district.[39] Transport was part of a commercialized landscape that moved locally produced or refined goods to a wider market.

For all the economic importance of manufactures and shipping to the local economy, a bona fide city needed a commercial base to thrive. The earlier failure of the Ebenezer Allan mill site had proven that the potential of the Genesee Falls site was not immune to poor planning. Both industry and commerce were credited to the city's success. By 1819 one traveler surmised that in Rochester both "commerce and manufactures are carried on with the facility and steadiness of a Hanse town."[40] At the 1834 legal incorporation of Rochester the new mayor, Jonathan Child, tapped directly into the rhetoric of improvement, noting its commercial and industrial transformation from an unimproved landscape into a righteous and productive one. Rochester "has been settled and built, for the most part, by mechanics and merchants, whose capital was economy, industry, and perseverance. It is their labour and skill which has converted a wilderness into a city."[41] Decades of this kind of city building permitted the 1880 United States Census summarily to conclude that "the very existence of a city indicates the presence of manufacturing and commercial enterprises."[42]

Designating the commercial district and separating it from residential purposes required more absolute controls and subtle clues than did defining the milling district. On one level there was a certain acceptable vagueness about the commercial-residential distinctions. When designing his city, Nathaniel Rochester specifically mentioned only "mill lots" and "town lots." Upon the town lots, most of which averaged a quarter-acre, he stipulated in the sale contracts that either a dwelling or store be constructed within a short period.[43] In an age when shop houses were a common building type and could be found in neighborhoods that were either primarily commercial or residential, the general appellation *town lots* seems appropriately flexible and yet contrary to the sorting impulse.[44]

The city planners did not shape the residential landscape as a de facto district. While many historians have looked to the separation of work and home as the first step of sorting, the pattern displayed in the new nineteenth-century city indicates that other forms of sorting preceded the division between the domestic and occupational spheres. In the new cities residential districts developed as an afterthought in the formal planning process, even though they consumed the bulk of the total acreage. Neither Nathaniel Rochester nor Joshua Forman specifically designated residential neighborhoods in their urban plans. Public, not private, urban spaces dominated their attention. Residential areas were seen more as the space left over after the industrial, commercial, and civic zones were demarcated, rather than as a critical zone of attention.

Private, residential areas were of secondary importance in the public, urban venture of staking out and promoting a city. Sorting the public cityscape preceded sorting the private one. Although not the focus of this study, it is worth noting that elite residents in the sorted city did show a preference for class-based residential sorting. Within a generation of settlement, residents established elite versus working-class residential pockets. For example, Rochester's rarefied Livingston Park of the 1830s was clearly more privileged than immigrant-tinged Dublin by the Falls. Conveniently located immediately south of the commercial district, Livingston Park featured stylish, large houses built by the native-born, Protestant social elites of the city. In contrast, Dublin by the Falls developed on the opposite side of the Genesee River, in closely spaced houses occupied by the Irish Catholic working class. Yet even this separation was not quite that clear. In the fashionable neighborhood of Livingston Park the neighborhood's secondary streets hosted less prominent families and even working-class whites and free blacks.

Despite the fluidity of commercial and residential options implicit in Nathaniel Rochester's broad category of town lots, city founders coaxed commerce into a particular district. By manipulating street widths, lot orientations, and lot pricing, they created on paper the appearances of a central business district at Rochester's Four Corners and Syracuse's Clinton Square. Wide streets flanked by narrow lots suggested busy street arteries with high land values, the essence of a commercial district. Wide lots facing narrow streets suggested lower land values and less trafficked secondary streets—that is, a residential district. Both Rochester and Syracuse were platted with these truisms in mind. At a time when the state ruled that primary roads must be four rods (66 feet) wide, the thoroughfares crossing at the Four Corners in Rochester and Clinton Square in Syracuse were a bold six rods (99 feet) wide, and the secondary streets laid out a proud four rods wide. The typical quarter-acre lots that faced the "main" streets had their narrower, four-rod frontage to the street and ran back ten rods (165 feet). In theory narrow frontages reflected the increased competition and value of frontages on the main streets, but these narrow urban lots were laid out in anticipation of competition, not as a result of it. This pattern, associated with the rise of the central business district, predated any real business.

Price differentials also abetted the sorting of urban districts. Although symbolically satisfying wide streets shaved salable land from the market, the monetary loss was compensated by the promotional value of a major highway and the economic bonus of the value added to its facing lots. Nathaniel Rochester charged thirty dollars for lots on the less important narrower streets but collected fifty dollars for the same sized lots on the primary streets and even mustered two hundred dollars for the half-acre northwest lot at the Four Corners.[45] By charging more for lots in the immediate compass of the Four Corners, Rochester introduced preconditions designed to make investors put the more expensive lots toward remunerative, commercial use.[46] If revenue-producing enterprises such as shops, stores, and inns lined the widest streets, the founders' plans and the commercial district would be realized. This scenario played out perfectly at lot No. 1. Its first purchaser, Henry Skinner, constructed a log cabin in 1812 and promptly turned it to remunerative use, renting it out to the Scrantoms, the first family of the village. Although the house may not have been particularly "commercial," it did produce both revenues and settlers at a critical time in the city's development. Just five years later Skinner sold the $200 lot for $11,200 to the Ensworth brothers, who moved the log cabin to the back of the lot and opened up the Ensworth House. The successor Eagle Tavern and Hotel remained one of the premier institutions in the economic and social life of the city until it, too, was replaced in the 1860s by the still-extant, dramatic pile of the Second Empire Powers Block (Figs. 30–32).[47]

Competition for the Four Corner's commercial lots indicates that purchasers literally bought into the idea of the center's heightened utility and value.[48] One of the earliest settlers, Abelard Reynolds of Massachusetts, was stymied in his first attempts to buy a lot in the unsettled city in 1812. Having caught what he called "a touch of the western fever," he found himself in the wilderness of Rochester, "which had just been mapped as a village covered with the native forests."[49] With the only evidence of a city being the plat in his hand and a land agent by his side, Reynolds pointed to several corner lots, only to be told they had already been taken, including the largest, $200 corner lot, which had gone first. Turning down suggested lots on the industrially inflected Mill Street, the commercially driven Reynolds requested two lots on Buffalo Street lying between the Four Corners and the river, only to be told that they belonged to the agent. After a short discussion, the agent agreed to release his claim, and Reynolds purchased these double lots on which he built his house and opened his saddlery shop, tavern, and post office. In 1826 Reynolds began an ambitious building program that reconfigured his primitive shop house into an elegant arcade, and he relocated his residence to Sophia Street (see chap. 6).[50] At this earliest stage of city building Nathaniel Rochester's sorted template had already shown its utility: the paper articulation of platted districts led Reynolds to avoid the future mill and warehouse district, which held no professional interest for him. Of course, that same mill district also gave Reynolds reason to believe the mill village could become a market town.

Rochester's sorted cityscape included a small civic pocket bordering Buffalo and Fitzhugh streets, one block from the Four Corners. Optimistically reserved for a future county seat, this civic function played directly into the rhetoric and practice of commercial urbanism. County seats were good for business, and cities competed for this advantage. It was not simply that being the county seat gave status, although a pretentious courthouse and coterie of professional circuit court lawyers certainly could add luster and respectability to what could be an architecturally and socially unpolished place.[51] County seats brought people into town. The sure and steady work of county administration meant a reliable stream of people with court business, all of whom needed urban services and spaces. Hotels offered accommodations, food, meeting rooms, and livery stables. Shops provided an opportunity to make necessary or indulgent purchases.

Nonetheless, the way in which Nathaniel Rochester platted the county reserve indicates the consistent subordination of civic activities to commercial ones in his sorted city. The term *subordination* is especially appropriate, for it implies a relationship, albeit hierarchical, as opposed to separateness. The county courthouse played a supporting role in the performance of the commercial district, tucked in as a pocket at the outer edge of the commercial and industrial quadrants. Nathaniel Rochester was willing to give up a bit of space—but not too much or too valuable a space—for the hoped-for eventuality of winning county seat status. He had initially planned to reserve the prominent lot No. 1 at the Four Corners for the county seat but quickly changed his mind and bumped the courthouse reserve to the southwest margins of the central district and kept the large corner lot for sale. Although a courthouse was a valuable urban building block, it did not need to be in the middle of town in order to be promotionally or functionally valuable. In fact, positioning it at the Four Corners would have usurped the commercial utility of salable, improvable ground and perhaps diluted the message about the city's central focus on commerce and profit maximization. The city was doubly rewarded. Lot No. 1 became an immediate and consistent showcase of commercial vigor, and Rochester became the county seat of the newly formed Monroe County in 1821. That close but peripheral civic lot became a star in the urban and economic constellation of Rochester, featuring not only the 1822 Monroe County courthouse but also the 1824 Presbyterian church behind it and the 1823 Episcopal church and the first school across from it on Fitzhugh Street, all constructed on lots donated by Nathaniel Rochester.

Positioning the courthouse at the edge of the economic center of a city, where it augmented commerce without usurping its focus, also occurred in Syracuse. Unlike Nathaniel Rochester, Joshua Forman was not as politically ambitious in his plans for Syracuse and never inserted a county seat space in his plat. Perhaps he erroneously believed the closeness of the existing county seat at Onondaga Hill, only four miles away, made future changes unlikely. Slow to gain county seat status, Syracuse's eventual 1856 Onondaga County courthouse did not command its own honorific, articulated square symbolizing civic importance. However, like the Monroe County courthouse, it was tucked into a pocket at the edge of the commercial district. There on the northwest corner of Clinton Square its forecourt was the existing markets and docks of the Erie Canal.

The sorted city can be seen as a city run on business principles. The platting of a sorted cityscape expressed a commercial urbanism, which was itself intended to appeal to members of the entrepreneurial merchant class who would then assume responsibility for carrying out that message as they built up the city. Rochester's sorted, efficient, economic rationalization of space reiterated commercial opportunities in each of the three districts. The commercial district was prominent and its trade connections clear and convenient. The county seat reserve promoted further business opportunities. The milling district promised a few jobs, a cash market for Genesee wheat, and money in the pocket of the farmers passing through the commercial district. Sorting also enhanced each district's internal efficiency by discouraging non-like activities. Shops clustered for maximum impact and viability. The courthouse was convenient to business without interfering with commercial space and activities. The millrace extended only as far as it was needed and no farther. The cityscape envisioned by the founders was that of an economically productive bundle of districts bound together into a commercially oriented whole.

Compactness

Nathaniel Rochester not only delineated three separate functions within his city but did so within four tight blocks. Joshua Forman, too, understood the necessity of compactness in a new settlement even before he began his Syracuse project. Having moved to the new village of Onondaga Hollow in 1800, he "began early to manifest his public spirit and enterprise" by "building up the village, and of extending its boundaries."[52] In his first project of urban boosterism, using personal and outside investment, he forced into being a hotel, academy, church, and stores in the center of Onondaga

Hollow with residences at the margins, "thereby connecting the whole into one tolerably compact settlement."⁵³

The founders of these new cities believed that a territorially compact city would more likely ensure commercial success. William Cooper, the founder of Cooperstown, New York, explained in his 1810 *Guide to the Wilderness*, "I hold it essential to the progress of a trading town, that it be settled quickly and compactly."⁵⁴ *Quick* and *compact* were actually related in promoting development. Compactness involved small lots that cost little and left the settler some money for actual construction. Compact layouts also provided a sense of density and bustling productivity which would attract possible investors or settlers.

The very language of Cooper's rhetoric of compact town planning targeted the commercial interests dominating city building in New York State: "The labor of two or three hundred industrious men concentrated, is like money collected in a bank; when scattered in distant quarters its effects amount to little; when brought together it resembles the heart, from and to which circulation flows, whilst it gives life and health to the remotest extremes."⁵⁵ The close proximity of people, goods, services, information, and ideas that informed Cooper's circulatory metaphor is what the geographer Allan Pred calls the "multiplier effect," in which the sum of produced and transmitted values becomes greater than the parts.⁵⁶ The effect, of course, was not just that the city's own economy prospered but that the health of the entire economic system in which the city operated did as well. Compactness had been elevated to a capitalist tool in the market economy.

Cooper's directives focused on a commercial urbanism. "Villages and trading towns built on extensive lots, where the inhabitants are dispersed, never make much progress in trade," he explained.⁵⁷ Small lots forced the economic life of the settlers toward an urban rather than village economy. Small lots ruled out farming and forced a settler to urban pursuits. If large lots were incorporated into the plan, Cooper warned, the city would not thrive:

> [If the mechanic] devotes his time in any manner to the cultivation of the ground, even to the raising what is sufficient for his own consumption, it is impossible he should be so active and vigilant in his profession as he would otherwise be; for instance, the barber, instead of being on the way to shave his customers, will be found weeding his onions; the shoemaker hoeing his potatoes; the watchmaker, who has neglected to repair the farmer's watch, will think it an excuse that his orchard required his immediate care; the blacksmith, who has been tilling the ground . . . will be unable to shoe

the traveler's horse, or mend the carrier's wagon. . . . if there be another village or market, though more inconvenient or more distant, where every man is to be found at his post, the mechanic in his workshop, the merchant in his store, such a place will carry off the prize of industry, and become the rendezvous of traffic and prosperity.[58]

Cooper believed that the town should be dependent on the country, and vice versa. Small lots eliminated distractions, including any lure of self-sufficiency, which would compete with business.

Although Cooper did not specify the exact layout of a compact plan with small lots, he seems to have favored a grid, having used it himself in Cooperstown.[59] The founders of Rochester and Syracuse apparently agreed with Cooper's principles as they laid out compact settlements of approximately quarter-acre lots set within irregular grids. They also encouraged dense settlement and discouraged pure speculation by restricting the availability of lots on the market at any one time. City founders regulated the supply of lots in order to maximize the effects that demand would have. Nathaniel Rochester sold only central lots at the Four Corners before opening up lots farther out. Limiting the number of lots left few options but to purchase in the center, creating the density Cooper advocated. It also raised the price of lots, a result that rewarded Nathaniel Rochester and his partners with more initial sales revenue and also, not coincidentally, urged purchasers to put the lots to remunerative use. Just as Nathaniel Rochester staggered the timing of bringing lots for sale, so, too, did the Syracuse Company, the largest landholder in Syracuse after 1824. As late as 1834, John Townsend, a key administrator of the Syracuse Company, still resisted selling a portion of the blocks just north of the canal, correctly believing that manipulated scarcity not only would enhance the value of those lots once they went on the market but would also maintain the current focus on developing the city center.[60]

Quick construction in these new cities made them seem even more dense and prosperous. In fact, the founders of Syracuse and Rochester pressed residents to build rapidly to transform the place from a city on paper to a city of wood and brick. Actual construction should make the city appear economically robust and help avoid mere paper speculation in land values. As the constant increase in land prices indicated, simply staking out a city had enormously added value to the site. The value of Rochester's 100–Acre Tract and Syracuse's Walton Tract rose through the years, but the speculative bubble threatened to burst unless settlers improved the land and justified the rising prices. Thus, Nathaniel Rochester's contracts required owners to construct a building "not less than twenty by sixteen feet" on a town lot within a year or the property would revert to the sellers.[61]

Compactness had promotional value, always a bonus for new settlements. When the New York City resident Johnston Verplanck passed through Rochester in 1822, he poked fun at the tree stump–pocked streets, yet he concluded that this was no backwater village because of the density of construction. The "houses are built in a compact manner, indeed more so than is necessary in a country town. The inhabitants already are anticipating incorporation as a city, at no late period, and in all probability their wishes will not be disappointed."[62]

Compactness also suited the genteel sensibilities of the merchant-led middle class, which was leading the commercial urbanization of New York State and trying to construct a spatial culture that valued gentility.[63] Cooper himself believed that compact building improved the tenor of frontier society. Urban compactness was the opposite of rural isolation. Close neighborly contact, he argued, inhibited the slovenly habits that otherwise multiply in privacy: "A kind of city pride arises, and acts advantageously upon the manners and modes of life; better houses are built, more comforts introduced, and there is more civility and civilization."[64] Cooper, like many champions of westward urbanization, promoted cities as a moral good that advanced civilization.

Intentions

The founders' first plans were idealized visions of their intent to build mercantile cities through the use of a sorted cityscape. Both the commercial goals and the sorting used to achieve those goals were still quite fluid at this point. In reality the broad streets and jostling lots were merely theoretical lines running through forests, fields, and swamps. Even the regularity of the inked plan, according to the urban geographer James Vance, "was orderly only on the first map filed in the county courthouse."[65] Nonetheless, the initial decisions of the developers of Rochester and Syracuse influenced the course and form of each settlement's advance. Early decisions were important. Using the tools of a physical plan, which included transportation routes, grid plans, lot size, orientation, pricing, and marketing strategies, the developers provided a two-dimensional armature for a three-dimensional sorted cityscape. A plat could not concretize a paper village, but it could, and did, suggest the shape of things to come. The cultural geographer William Wyckoff has described a developer's legacy on the landscape as contributing the "essential skeleton" upon which the settlers hung the "flesh and fiber."[66]

The initial surveys of streets and lots and designations of particular districts resulted in important and durable configurations in the new nineteenth-century settlement. The physical form and specialization of each city's districts would evolve over the decades, but the essential characteristics were put in place during the first stage of settlement. This land assignment was durable; the armature of the settlement laid the groundwork for subsequent improvements. Created in advance of actual settlement, the plans by the founders showed a keen understanding of the impact that urban form could have on generating enthusiasm and investment in the cities. The early decisions specifically showed a preference for functionally sorted cityscapes that abetted a particularly commercial urbanism.

City planning was one thing, but city building would be the proof of an urban vision. Animating the plat with buildings and people would be the truest test of the spatial culture of sorting. Clearly, there was interest in urbanizing the eastern frontier. Unsolicited queries about future land sales had emboldened the proprietors to envision an urban venture greater than the paltry mill seat that had first drawn them to the tract. "I have been applied to for building lots there," a pleased Nathaniel Rochester wrote to Charles Carroll on January 13, 1811, before they even had clear title to the land, "and there is no doubt of there soon being a village there and much business done if lots would be had."[67] It did not take long for them to launch the urban speculation. On November 19, 1811, just one day after the Maryland partners' title to the 100–Acre Tract was officially clear, Rochester reported to Carroll that he had already laid out about twenty-five lots, sold some of them, and leased a mill seat.[68] Building the sorted city had begun in earnest.

THREE

Building the Sorted City
The Three Epitome Districts

During the 1820s the English couple Basil and Margaret Hall were amazed by the paradoxical primitiveness and achievement of the new cities they visited in the New York frontier. "Rochester is the best place we have yet seen for giving strangers an idea of the newness of this country," decided Margaret.[1] Basil detailed the contradictions between Rochester's plat and the actual settlement. "Several streets were nearly finished, but had not as yet received their names; and many others were in the reverse predicament, being named, but not commenced."[2] He proceeded to recount the speed and rawness of the new settlement:

> Our friendly guide who was showing off the curiosities of the place, and was quite glad, he said, to have this opportunity of exhibiting the very first step in the process of town-making. After a zig-zag scramble amongst trees ... we came to a spot where three or four men were employed in clearing out a street, as they declared, though any thing more unlike a street could not well be conceived. Nevertheless, the ground in question certainly formed part of the plan of the town. It had been chalked out by the surveyor's stakes, and some speculators having taken up the lots for immediate building, of course found it necessary to open a street. ... As fast as the trees were cut down, they were stripped of their branches and drawn off by oxen, sawed into planks, or otherwise fashioned to the purposes of building without one moment's delay. There was little of exaggeration, therefore, in supposing with our friend, that the same fir which might be waving about in full life and vigour in the morning, should be cut down, dragged into daylight, squared, framed, and before night, be hoisted up to make a beam or rafter to some tavern, or factory, or store, at the corner of a street, which twenty-four hours before had existed only on paper.[3]

It was as if neither streets nor buildings could keep apace with the speed of urbanization: "In many of these buildings the people were at work below stairs, while at the top the carpenters were busy nailing on the planks of the roof. . . . we saw great warehouses, without window sashes, but half filled with goods."[4] The process of physically building a city was a noisy and messy effort that belied the orderly planning behind it. Basil further noted the "fine concert" of Rochester in action, "half-finished, whole-finished, and embryo streets were crowded with people, carts, stages, cattle, pigs . . . all these were lifting up their voices together, in keeping with the clatter of hammers, the ringing of axes, and the creaking of machinery."[5]

Yet, even within the primitive conditions of this rapid and rough-hewn city, observers could discern the compact and sorted cityscape. As early as 1815, a state surveyor's sketch noted a compact community clustered at Rochester's Four Corners (Fig. 6). Moreover, the arriving merchants and millers had largely adopted the city founders' mercantilist visions as their own, both in terms of the physical sorting of the cityscape and its intended message of productive opportunity. Two illustrations made of Rochester during the 1820s showed the definite districts that were geographically, functionally, and architecturally distinguishable (Figs. 8–9). Within two blocks of the Four Corners, the three hallmark districts each boasted their own defining identity that was immediately legible to even the most casual observer. During the first stage of actual construction the settlers created a legibly sorted cityscape through the manipulation of a building's location, function, and, occasionally, style.

While on his American tour, Basil Hall sketched the south side of Buffalo Street from the vantage point of his room at the Eagle Hotel, the brick building situated on that prominent lot No. 1 at the Four Corners (Fig. 8). Casting his sights slightly westward, his sketch captured the crude but compelling physical and architectural distinctions between the commercial and the civic districts. Public Square Alley split the view down the middle, with the commercial district of shops to the east and the civic district with the courthouse and churches to the west. The view was a study of contrasts: frame versus masonry materials, domestic versus monumental scales, vernacular versus high-style references, low versus high profiles, and simple stoopless facades abutting the street versus elegant set-back terraced porticoes.

The commercial aspect was clear, if not elegant. Buffalo Street was a rutted morass without sidewalks, but the contiguous rows of buildings clearly distinguished it from a street in a rural village. In scale and material the frame two-story structures were not unlike many dwelling houses of the day, but the multipaned, large windows articulated a commercial function where light and display were important. Painted signboards broadcast the commercial and artisanal enterprises of Wakelee's boot shop, Russell Green's painting and glazing shop, Landfear's cabinetmaking shop, and Ambrose Chapin's grocery. The contiguous construction of cheek-by-jowl building indicated that Nathaniel Rochester's dreams of competition for main street lots were coming true, building by building. A workshop upstairs and provision store in the cellar further registered the intensity of land use in the commercial district. Nathaniel Rochester's manipulated scarcity of central lots, the high price of those lots, and the convenience of a position along the state road to Buffalo all persuaded merchants to develop their commercial enterprises here.

The civic aspect was altogether different, boasting architectural elegancies not found in the commercial district. The three-story 1822 Federal-style Monroe County courthouse sat back from the street and was raised upon a terrace, giving the building particular prominence and, not coincidentally, elevating it above the swale that had previously held a frog pond. Its stone construction, Ionic portico, gable-end fanlights, and octagonal cupola proclaimed its status as an important public building. Its neighbors were other similarly high-style buildings of institutional import. The small Greek Revival building held professional offices that were shortly converted to the court offices. Two stone churches shouldered Fitzhugh Street: the Presbyterian church with its Neoclassical spire rose from behind the courthouse, and the Gothic Revival turret of the Episcopalian St. Luke's peeked over the cornice of the courthouse. The proliferation of cupola, steeple, and tower all gave the civic district a skyline presence, a feature lacking in the lowly commercial district.

The German traveler Adolphus Duttenhofer made his sketch one block away (Fig. 9). An engineer touring America's canals, Duttenhofer sketched the intersecting shipping and milling district in Rochester's southeast quadrant. The stone aqueduct carried the still water of the Erie Canal over the turbulent Genesee River. At 802 feet long with eleven 50–foot arches, the aqueduct itself was a major technical feat by the engineer Benjamin Wright. Mule-towed canal boats, perhaps bearing wheat, headed into the city, past several mills. The slip known as Child's Basin held a docked boat not far from the 1821 Red Mill and warehouse that straddled the millrace. Sturdy, gable-roofed mills with their waterwheels cutting into the river and millrace stood at the ready, their dangling hoists prepared to raise raw wheat to the attic where its milling began. A narrow footbridge, famously low as were most bridges along the canal, connected the town on either side of the canal. Having just passed under one, passengers scrambled to their feet on the boat's roof deck. The emphasis in this industrial district was on water for power and transport. It was an economically, functionally, and visually unique landscape in the city, physically close but functionally and visually far from the commercial or civic districts sketched by Hall.

Commercial District

The initial plattings by the city founders correctly anticipated and fostered a commercial district at the nexus of major crossings. Here, at the transportation hubs of the new cities, retailers, wholesalers, craftsmen, and professionals could conveniently connect with local and distant customers and were well positioned to import and export goods.

Buffalo Street in Rochester quickly emerged as the primary commercial street, seconded only by its cross streets of Carroll (to the north) and Mill (to the south). By 1818 land agents were promoting their tracts in terms of proximity to Buffalo Street and the bridge. Buffalo Street was so well situated for business that even the Genesee River could not interrupt the flow of commerce. In 1821 a Rochester newlywed happily reported how business was booming in her adopted home, writing that one could hardly cross the Buffalo Bridge for all the traffic in the city.[6] Another traveler that year marveled that "the streets, which are spacious, present a succession of well-furnished shops, and the bustle which continually pervades them, gives the whole place an air of activity and commerce."[7] In 1824 the original plank bridge was rebuilt in stone and bedecked with shops, including the market house, a grocer, shoe dealer, bookseller, and drugstore, built directly along its sides in medieval fashion.[8] Entrepreneurs had transformed Nathaniel Rochester's wide streets into an incontrovertible commercial zone, and the first city directory of 1827 recorded more businesses on Buffalo Street than on any other.[9]

Syracuse similarly developed an articulated commercial district containing an array of businesses. By the 1820s the pioneering Bogardus Inn, which had been contracted into existence according to the state's sales terms, had been upgraded to the Mansion House and augmented by neighboring shop houses whose owners agreed with the commercial potential of the transport-laced junction (Fig. 7). The frame buildings housed a commercial mix of artisans, retail and wholesale merchants, and professionals, including a shoemaker, hardware dealer, hatter, druggist, jeweler, grocer, saddler, barber, bookseller, dry goods merchant, hide and oil dealer, lawyer, grain loft, and residences.[10]

The mix of retail and wholesale businesses next to the turnpike and canal carried its own logic and symbolism that aided a new settlement. Relying on face-to-face transactions made over the counter in small quantities, retailers needed a convenient location for shopping and other neighboring shops to attract customers. Dealing in bulk goods at large scale, wholesalers needed to be close to efficient and reliable forms of transportation. In pioneer regions with immature commercial centers, merchants typically relied on both the local retail and regional wholesale market for support.[11] The Buffalo Street "Rochester Cash Store," a wholesale and retail dry goods and grocery store thus advertised itself in 1827 as dealing in the "goods as are usually wanted in 'Town and Country.'"[12] Transportation junctions offered advantages to both commercial segments.

Clustering became a self-fulfilling prophecy. As merchants and artisans recognized the benefits of a central location, they flocked to the locale, and each subsequent addition confirmed the sagacity of the previous one. Competition for prime lots confirmed the importance of location. A new hardware merchant assessing the competition in Syracuse paid for a new survey that, not coincidentally, corrected a platting error by extending a new corner lot at Salina and Genesee streets. He promptly bought the new lot and erected a brick building that dwarfed his competitor's hardware shop on the old corner lot.[13] The incident well illustrates the process of shaping a commercial district: consensus on the best locations, rivalry to command advantages, and contiguous buildings whose plan and height profitably and prominently maximized their productivity. Success drove businesses to expand the commercial district's territorial bounds within a few years. With Salina three miles north and the canal towpath also positioned on the north side of the canal, Syracuse merchants anticipated more activity toward the north side of Clinton Square and a northerly growth of the commercial district. One newcomer defended his family's decision to buy a lot north of the canal despite its greater cost because "few persons, if any, then thought that the south side of the canal would ever be anything."[14] Similarly, when John Wilkinson bucked expectations and invested in the south side, opening his law office on South Salina and Washington streets in 1819, he was "heartily ridiculed for putting his office out in the fields." The site was two short blocks south of Clinton Square yet was considered "then out of town."[15]

The density of party wall businesses was the primary indicator of a commercial district, but architectural details also distinguished the commercial purpose of the buildings. The density of commercial activity on narrow lots translated into rows of adjacent, or party wall, structures. Although there were similarities with domestic architecture, the combination of close spacing, small lots, signboards, display windows, hoists, loading docks, balconies, and pumped-up scale pointed to a commercial use. Shops with their gable ends turned to the street were another commercial clue. While on a scientific reconnoiter of western New York in 1819, one traveler was surprised that Rochester already displayed a commercial streetscape, writing that Buffalo Street, with its *"stretch* [of] a mile or more," is "covered with houses and opulent establishments," and these stores, "with their gables to the street, are shewy [sic] and well stocked."[16] Buildings with their narrow gable ends to the street reiterated the narrow and deep configurations of their lots, which themselves had been platted to suggest competition for central lots.

Although the attitude would change during the second wave of settlement, at this early stage the location and stock of a business were more important than its outer shell or style. With a few notable exceptions, shopkeepers served a captive retail customer base with little mobility or a distant wholesale customer who cared little for architectural amenities. Nor did shopkeepers yet feel compelled to upgrade their buildings for individual promotional or civic pride reasons. During the first stage the commercial streetscape remained one of mixed building materials, mixed commercial activities, and mixed degrees of architectural refinement. At the same time it was also sorted, displaying a strong commercial purpose and identity.

Civic District

A recurring theme in the mercantile city is the subordination of civic interests to the commercial. Responding to a cultural climate that stressed economic productivity, the founders and settlers devoted their city-building resources and energies toward business and industry. Spatially, this attitude played itself out in the ranked hierarchy of districts within the city center, with commerce and industry given primary prominence over a secondary—but still important—civic node. City builders knew that a county courthouse was an important place-making device, so it was important to have it spatially and architecturally visible in a competitive city. But the size of the courthouse square and its precise location directly reflected its supporting, as opposed to starring, role in the mercantile city. Given its proportionally smaller size, it may be more appropriate to consider the civic district more of a civic pocket.

The courthouse was a vital place-making institution that conferred political stature, boosted local businesses, and gave a competitive advantage in attracting settlers. But committing to a courthouse required balancing several competing agendas. Public visibility and user convenience pointed toward a large, expensive, prominent downtown site. Civic values suggested giving the building an honorific setting, where its luster could polish both the social and architectural tenor of the city. But in a compact city a grand spatial gesture downtown would remove profitable lots from market. The compromise in the commercially oriented new city was to locate the courthouse adjacent to, but not in, the commercial district. This geographic compromise is the locational strategy Nathaniel Rochester pursued in 1811 and the one the Syracuse city council followed in 1856.

Although Nathaniel Rochester correctly anticipated winning county seat status, the ten-year lag between the platting of the city and the state legislature's approval of a new county seat forced local authorities to reconsider the actual location of the courthouse. During that decade the city of Rochester had grown not only by infilling the original 100–Acre Tract but also by absorbing rival settlements around it. In 1817 it annexed Frankfort, with its powerful Middle Falls to the north, and in 1823 Rochester officially crossed the river by annexing East Rochester at the far end of the Buffalo Street bridge. The annexations strengthened the case for Rochester deserving county seat status but also complicated the issue of where to situate the courthouse.

The first issue to resolve was that of the county designation. Although the Buffalo Street bridge linked the original 100–Acre Tract on the west side and the spillover settlements on the east side, the Genesee River legally separated Rochester into two different counties. Legal business relating to the original Rochester on the west bank of the river needed to be handled thirty miles southwest, in Batavia, the Genesee County seat. Legal issues pertaining to the east bank had to be conducted twenty-five miles southeast, in Canandaigua, the Ontario County seat. The inconvenience of a bifurcated county jurisdiction was compounded by the fact that Rochester was the fastest-growing and largest city in either county and thus was disproportionately burdened by the arrangement. Led by Nathaniel Rochester, settlers who lived in the environs of Rochester petitioned the state legislature to create a new county with a new county seat, arguing that, just as Rochester was their "natural market" for ordinary business, it was also their natural center for legal business.

Conducting legal business elsewhere not only inconvenienced the public but also disadvantaged the local economy. Many residents believed that their subordinate role in the counties stymied the city's growth to the return advantage of Batavia or Canandaigua. At least one Rochesterian argued that businessmen in Canandaigua and Batavia were actually the city's opponents, who viewed the upstart settlement "as an intruding stranger, who had seated herself too near them, and wished her strangled."[17] There is some credence to this argument, given that Batavia and Canandaigua interests had protested the extension of the state road through Rochester, had perhaps even been complicit in the attempted diverting of the road, and had certainly campaigned against the state chartering banks in Rochester. But the booster Nathaniel Rochester had demographics on his side. The number of people who would benefit by the creation of a new county outweighed those who would benefit by stasis. In 1821 the state legislature finally capitulated and carved out a new county.[18] Anticipating an improved home life, the Rochester wife of a lawyer who rode the circuit court happily wrote to her husband that his days of travel were over: "the petition for a new County, which has been presented several winters is at last granted & you may expect soon to hear of our courthouse and jail, the name of the county is to be Monroe."[19] The courthouse played an important role in the urban identification of Rochester. Prior to 1822 Rochester was formally known as "Rochesterville"; upon completion of the courthouse, however, residents in the new county seat officially dropped the parochial *ville* in favor of more dignified *Rochester*.[20]

By the time the new county was approved, there were several potential sites for the courthouse in a territorially expanding Rochester. Prior to being annexed into Rochester, developers of adjacent tracts had also impressed courthouse aspirations into their plats. In addition to Nathaniel Rochester's reserved lot at Buffalo and Fitzhugh streets, there was the large central square in Francis and Matthew Brown's Frankfort tract which was intersected by a street optimistically called Court Street. Across the river in Elisha Johnson's East Rochester tract was yet another hopeful Court Street edging a center square (Fig. 3). Despite annexation, deeply rooted factions and loyalties divided the tracts. Prior to becoming the county seat, Rochester had faced its own internal politicking around its village charter. In the first village trustee elections of 1817, lines were drawn between the merchants and lawyers associated with Nathaniel Rochester and the 100–Acre Tract (Bucktail Republicans) and the merchants and manufacturers associated with Francis Brown and the Frankfort Tract (Clintonian Federalists).[21] In both the trustee elections and the courthouse location, the Nathaniel Rochester faction carried the day. Although the Frankfort and

East Rochester squares were urbanistically far more elegant than the remaindered, half-block county lot Nathaniel Rochester had reserved, their distance from the commercial district satisfied neither the geographic structuring of court and city business nor the promotional posturing that a prominent courthouse along the main thoroughfare offered. The courthouse was not to be an institution isolated from the economic, social, or political life of the city, and its location reflected that integration. The principles of compactness and sorting which had guided the platting of Rochester still held true in building the districts and in the institutional relationships among them.[22]

Architectural style and proxemics reinforced the civic identity of this block on the margins of the commercial district. An 1837 view clearly articulated the civic pocket that had been suggested by Basil Hall's sketch a decade earlier (Figs. 8, 10). The 1822 Federal-style courthouse, with its Ionic portico, presided over its clustered civic subjects. A low fence and a raised, landscaped terrace separated the courthouse from the common street while making the courthouse simultaneously more prominent within the streetscape. A pair of Greek Revival temples in the Doric order, used for law and medical offices, bracketed the corners. The pinnacled and spired Neoclassical Presbyterian church rose behind the courthouse facing an interior square the two edifices shared. The Gothic Revival St. Luke's Episcopal Church and the public school stood across the way on Fitzhugh Street. Silas Cornell's 1839 map of Rochester included a vignette of the civic pocket as one of its few illustrations, another indication of the importance of this node in the physical and cognitive mapping of the city.[23]

In Onondaga County the location of the courthouse was extremely fluid, fluctuating according to a town's commercial fortunes within the county. It is indicative of Syracuse's long struggle to dominate the local political economy that it did not earn the courthouse until 1856, sixty-two years after the county had been formed and fifty-two years after the city was first surveyed. The peripatetic county seat began in Onondaga Hollow in 1794, the site of early white settlement and moved to nearby Onondaga Hill in 1801. After the Erie Canal stimulated business in salt-producing Salina and shipping-oriented Syracuse, the county commissioners decided in 1827 to move the courthouse to that emerging center of business.

But to which center? Whereas Salina was the dominant salt producer through the State Salt Reservation, Syracuse was the local leader in shipping activities by virtue of the Erie Canal. Both had genuine economic prosperity. The Syracuse Company, now the leading property holder in Syracuse, offered an entire block (with the exception of one lot given to the Presbyterian church) which lay two blocks south of Clinton Square, in a spatial pattern similar to Rochester with its civic pocket adjacent to the commercial center.[24] Salina had the weight of state authority in its founding and oversight of the saltworks. But the village trustees of Salina, doubting their clout and fearing Syracuse's rising importance, offered a free block on North Salina Street and the aptly named Division Street that lay in between the two rivals. Although inconvenient to Salina's business district, the advantage of this distant lot to Salina was that it kept the courthouse out of Syracuse. Seeking to avoid clear favoritism, the county supervisors accepted Salina's offer, and the compromise courthouse was built in the middle, equidistant to Salina and Syracuse (Fig. 5).[25] Architecturally, the compromise courthouse shared the classical vocabulary of power of the courthouse in Rochester. Completed in 1830, the compromise courthouse was brick, embellished with a columned portico, topped by a hipped roof, and surmounted by a cupola.

County seat status directly affected the host's urban fortunes. The loss of the courthouse weakened Onondaga Hill, which had enjoyed the county seat privilege for fifty years. The economic implications and the links to business were obvious. One Syracuse resident recalled, "one store or shop after another began to break up in the neighboring villages and establish itself here. In 1828, the county buildings were removed from Onondaga Hill.... This was the signal for a more general migration of business and professional men hither from the neighboring towns, and gradually, or rather, *rapidly,* the business of the country has been centering here."[26] But the compromise Salina-Syracuse courthouse failed to benefit greatly either of its compromised hosts. In this case, where county supervisors sacrificed urban compactness for political impartiality, residents experienced firsthand the imprudence of not building compactly. The compromise courthouse centered between two rival cities did not significantly stimulate development in the two cities or in the new center, and thus it failed as a commercial opportunity. It was joined by the fittingly named Center House hotel, sometime also known as the Half Way House, but annoyingly even the clerk's office was a mile away in Syracuse. As James Silk Buckingham surmised after his visit during the 1830s, the compromise, "like most compromises," satisfied no one except that both parties were equally inconvenienced.[27]

The compromise symbolized not the agreement but, rather, the tensions between two towns striving for urban prominence and economic dominance. Fights between Salina and Syracuse were quite literally fought on the courthouse grounds. An ongoing feud between the Syracuse and Salt Point "boys" drew both sides to what they called the "debatable ground," that midway point anchored by the courthouse. Here, one boy remembered, "we fout [sic] and fit, and gouged and bit, and struggled in the mud, until the ground for rods around was kivered with our blood."[28] It was hardly just boyish high jinks. On election days the courthouse on Division Street was darkened by clubs, stones, missiles, and barricades as the factions literally tore up the streets. Not just boys but also prominent men and active leaders were personally involved in these "knock down and drag out" melees.[29] In 1844 a riot in Syracuse between the two sides resulted in one man dead. Conciliatory minds called for consolidation under a city charter as a way to bring the two sides together. The more cynical ones called for consolidation in order to create a stronger police force that could quell riots. An agreement to incorporate as a city was finally reached when Salina received assurances that the new municipality would not strip Salina of its bank, salt reservation office, or post office.[30] In 1848 the merged villages of Salina and Syracuse were joined together as part of legal incorporation of the city of Syracuse. Erased as in independent entity, Salina was given the symbolic distinction of being Ward No. 1. William Leete Stone's prediction that Salina and Syracuse would be united "as Greenwich now is to New York" had come true.[31]

Legally united perhaps, but an arsonist's fire in 1856 at the courthouse cemented maneuvers that were already afoot to move the institution to Syracuse proper. Politics and economics joined again in the debate about where the new courthouse should be located. Unlike Rochester, where lots had been reserved from public sale by city founders, real estate in Syracuse had been in private hands for some time, which complicated the selection of a county lot. Like Rochester, the county and residents ultimately agreed on a civic pocket at the edge of the commercial district. All of the proposals placed the courthouse close to the commercial district, although just how close depended on the personal interests of the proposing landholder, who dangled free or discounted lots as the incentive. In the end the county supervisors accepted the offer of Colonel Vorhees, owner of the recently rebuilt Vorhees House (later known as the Empire House) on the site of the original Bogardus Inn and Mansion House hotel. Vorhees traded the city one of his lots on the northwest corner of Clinton Square in exchange for the burned-out old courthouse lot on Division Street.[32]

What was in it for Vorhees? He was less interested in the Division Street site than he was in having the courthouse as his neighbor. As the owner of the hotel and commercial block at the northeast corner of Clinton Square, he gained economically not only by robbing the Center House hotel of its raison d'être as the courthouse hotel but also by ensuring that his hotel was primely positioned to pick up the new business. Self-interest rightly understood made Vorhees both a savvy businessman and a generous urban booster.

What was in it for the city? Residents opposed to financing a civic square were conveniently enticed by offers of free lots by propertied merchants with their own economic interests at heart. The Syracuse builder Timothy Cheney had warned his country supervisors to "select the best site . . . with reference . . . to the conveniences of the people of the County" and not to be swayed by the interests of local property holders.[33] Local businessmen, lawyers, and county residents would have found a central location eminently more convenient. Vorhees's lot was also close to the site of the original county clerk's office on North Salina and Willow streets which had just been rebuilt in 1853. The location of the courthouse directly on commercial Clinton Square emphasized the interlocking relationship between the commercial and civic districts without really usurping much profitable land, since there was not extensive development out from the northwest corner. From its position on the northwestern edge of Clinton Square, the Onondaga County courthouse looked up Genesee Street to the nearby Baptist church, thus nodding to a civic and residential shift beyond the border. Not unlike the generation-older Rochester courthouse, this institution was also given a corner—but no more—of the commercial district. The residents considered the county courthouse primarily a tool of mercantile interests which rightly belonged in the business district.

Fashions had changed by the time Syracuse won its courthouse in 1856 but not the role of fashion. Syracuse followed the spirit of the Rochester example, whose 1822 Federal-style Monroe County courthouse was styled into a piece of contemporary, high-style architecture which sharpened the contrast between—and hence legibility of—the commercial and civic districts. The architect Horatio Nelson White designed the new Onondaga County courthouse in the latest Italianate style, with bracketed eaves and hooded windows and an eighty-foot-high tower. The picturesque style of the new Onondaga County courthouse, with its asymmetrical massing and detached form, made a dramatic, almost churchlike, counterpoint to the neighboring commercial rows of boxy, flat-fronted, shed-roofed buildings (Fig. 11). The difference between the commercial and civic architecture highlighted the differences between, and thus the presence of, the two nodes. The courthouse building committee argued with the county supervisors that the new courthouse must reflect well on the city and county: "The county of Onondaga is broad in territory, large in population, fertile in resources, and noted for its enterprising character. It must continue to advance with rapid pace in population and wealth. Would it be the part of wisdom for such a county to chaffer about the investment of a few thousand dollars in a building of this description? Should cheapness be substituted for durability and adaptation? The committee think not."[34] Built of Onondaga limestone, the courthouse put the very ground of the county on display. The stylish edifice also put the city on show. The building committee concluded that White's design trumped all the other county seats in the state: "[none] of them have a temple of justice more central, or more commanding in appearances."[35]

The Industrial District

The commercial district was the magnet that drew people into the city, and the civic district was a capstone to the city's political and economic achievements. The industrial district, however, was the backbone of the new cities' economies. Syracuse became known for its salt and Rochester for its flour. The two cities' industrial strength was often mentioned in the same breath, such as James Silk Buckingham's report that the Erie Canal made the new cities marts of business and "that what the grain and flour trade is doing for Rochester, the salt-trade appears to be accomplishing for Syracuse."[36] Industry in both cases meant not only the processing and warehousing of the region's raw resources but also shipping by exporting the local products and importing finished goods. As markets grew, the economies of both new cities expanded beyond the local hinterlands into regional markets and even national distribution, a perfect example of links within the great mercantile chain of new settlements on the eastern frontier.

The spatial distributions of milling and shipping were very site specific, and the architecture of industry was designed with an eye for interior function, not external display. Milling had to hunker in alongside the falls and millraces. Salt making needed to be close to the salt springs. The warehouses needed to be near the mills but also near the roads and the canal that would move out the refined products. Whereas the siting and construction of the commercial and civic districts weighed practical utility alongside symbolic considerations, the buildup of the industrial district was ruled by topographical realities. This is not to say that the industrial district lacked symbolic import—far from it. Travelers and residents alike uniformly pointed to industrial cityscapes as evidence of the land's opportunity and the citizens' enterprise. But the spatial delineation of the districts had an incontrovertible logic related foremost to production.

Capturing the waterpower of the Genesee River's fifty-seven-foot Lower Falls, the ninety-six-foot Middle Falls, and the twenty-foot Upper Falls had motivated millers since the turn of the century. The Upper Falls may have had the shortest drop, but the site enjoyed the converse natural advantage of the shallowest ford across the river, which ultimately gained Rochester the state road bridgehead and the Erie Canal aqueduct. This location of the waterfall and the lay of the derelict millrace anchored the southeast milling quadrant of the Four Corners. At the time Nathaniel Rochester and the first settlers were shaping the city, they operated on two assumptions: first, that there would be milling activity in Rochester proper; and, second, that Rochester's transportation advantages would permit it to transport the flour beyond its immediate hinterland. Mercantilism guided the founders' milling schemes as they envisioned a wider regional market opportunity from this local advantage.[37]

Correspondence between the Maryland proprietors made it clear that the mill site must be protected to preserve its industrial integrity as well as the reputation of the entire settlement.[38] City founders were convinced that securing industry protected the advance of the entire settlement. Upon hearing the news in 1817 of a destructive flood in Rochester, Charles Carroll immediately understood the deeper damage it wrought. The injury went beyond the harm to the mill site, or even the rest of the urban fabric, and extended to the entire reputation of the fledgling city. "We have already in public estimation sustained irreparable injury by the report of the destruction of the mills & the inundation of the Village," he wrote Nathaniel Rochester, "purchasers will look out for Scites [sic] not liable to those casualties."[39] To make matters worse, the Maryland partners were convinced that the flood was not entirely due to natural causes. The New Englanders who owned a rival mill settlement on the east bank of the river had erected a dam, which had the effect of channeling the river's overflow onto Rochester. Carroll accused the other side of perhaps even intentionally causing the flood. "The more we suffer in the eyes of the Public, the better for Brighton. I have learnt enough of Yankees to dread & fear their wiles & offers."[40] Even the flood, however, could not quell the appeal of the waterpower and transportation advantages, and the southeast quadrant of the Four Corners became an active milling center. The mill district, it seemed, was untouchable.

The annexation of Frankfort in 1817 and the routing of the Erie Canal through the original milling district during the 1820s altered the once sacrosanct mill district. The superior waterpower of Middle Falls was now linked to Rochester, which, coupled with the degradation of the old district because of the state's incision of the Erie Canal aqueduct and canal bed, emboldened many Rochester millers to relocate to the Frankfort Tract. Millers who remained in the old mill district, however, enjoyed the unrivaled privilege of the best route to market right outside their door. (Figs. 9, 12).

The architecture of milling was as distinct as that for the civic or commercial district. Just as display windows and signboards illustrated a shop's commercial purpose, the multistoried mill's ashlar walls, loading bays, hoists and pulleys, and tailraces diagrammed the production process. Just as a shop needed to face the street to serve and attract customers, the mill needed to connect with millrace and at least one avenue of transportation. Positioned along the millrace, gates regulated the water feed from the millrace, and sunken waterwheels turned the shafts that powered the machinery. The integration of mill and site awed an English industrialist:

> The flour mills are here on the most extensive scale of any in the Union, and the construction is superior to any in the world.... Every operation in these mills is more like clock-work than anything else; very few hands are employed, everything is done by water; the grain is hoisted up by power, carried to the smut mill, then to the hoppers, then to the stones, and from the stones to the cooling frames, and so on to the dressing or bolting mills, which latter are here preferred to the dressing mills of the old country; from the dressing mills it is shot into barrels, which when filled and weighed are immediately "ended up," branded, and ready for market. It is really pleasing to see such order and regularity in any manufactory; it is here done without bustle or hurry, and so clean and perfect, that the Rochester brand for flour stands pre-eminent in the markets of the whole world.[41]

By 1841 Rochester was milling about 500,000 barrels of flour per year, giving Rochester the appellation of the Flour City.[42]

The milling and transportation landscapes were avowedly pragmatic, built for purpose and not for show. And yet the scenario of the engineering marvels of the canal, aqueduct, and mills overlaid on a dramatic landscape of rapids and waterfalls struck contemporaries as a remarkable demonstration of man's improvement of nature. The public's fascination with the "industrial sublime" of the falls was fed by artists' powerful images that became part of the arsenal of the city-boosting endeavor.[43] The German engineer Duttenhofer described his own 1826 experience on the aqueduct in terms that mingled a romantic sensibility with industrial appreciation (Fig. 8): "I rode over it the first time under bright moonlight while the Genesee at that moment ran very strong.—Never has the contrast between art and wild nature struck me more strongly.—The same river which here approaches the falls so rapidly and unencumbered and which rips everything that it comes in contact with has to bear this bridge canal with its dead water so that the ships with their loads can glide over nature."[44] A more popular subject was the Middle Falls, whose awesome cataracts were paradoxically beyond human control yet harnessed for industry. An often reproduced view by W. H. Bartlett, published in Nathaniel Willis's *American Scenery* in 1840, codified the trope of the waterpower city (Fig. 14). The spray of the Middle Falls mists the view. Mills cling to the cliff edge. The commercial district rises in the background. Awed figures perch across the river drinking in the sight. The view is not just the sublimity of nature but also the clever hand of man in turning it to profit. Willis's description of the scene resonated with the ideas of the industrial sublime: "The descent to the lake is between the walls of a tremendous ravine, the grandeur of which seems to have had no terror for the souls of the manufacturers."[45]

The same engraving was reprinted in Charles Dana's 1854 *United States Illustrated* with a similar rhetoric of the sublime reconceived in industrial terms: "The interest which these Fall once inspired, has been entirely merged in the City of Rochester now surrounding their brink. We may say, indeed, that they have been swallowed up by the city, for while their waters have been mostly diverted for manufacturing purposes, the romantic feelings they once excited have been transferred to the unexampled growth of the town—far more a wonder than the natural object."[46] Whereas some observers were disappointed with the despoiling of nature, others, such as Frederick von Raumner accepted the combined drama of industry and nature: "The drawing off of part of the water from the first fall, for manufacturing purposes, has been censured as detrimental to its beauty; but I cannot coincide in this judgment. Without regard to its great utility, there is a romantic look in the situation of the buildings perched on the ledge of rocks, which from between and beneath them larger or smaller streams are seen plunging into the deep valley below."[47]

Written and illustrated views of the falls celebrated the very idea of improvement which guided the settlement of the interior. Upriver stood the commercial district built from merchant miller profits. Downriver the cliffs teemed with mills churning out the flour. The harnessed power of the falls tapped right into the cultural imperative of utility and productivity which girded urban settlement. Passing through Rochester in 1828, one traveler observed: "when its natural advantages are considered, I know no place which can compare with it. . . . It is calculated for as many mills as there are spots to place them, and the water can be used five or six times within the distance of a mile." It was with some amazement that he concluded, "water seems to be made to do everything here. The blacksmiths have become so lazy that they even make it blow their bellows."[48]

Public curiosity was generally indulged by welcoming merchant millers. Looking to experience the full sublimity of falls, the intrepid Fanny Trollope was "handed" from the doorway of one of the mills for a better view. Twenty years later Frederika von Bremer had a similar experience. She emerged dusty with flour from her "top to bottom" tour of a Genesee Falls flour mill.[49] Other visitors were more focused on the mill operations. "Through the kindness of the proprietor," the Canadian Thomas Rolph recorded in 1836, "I had an opportunity of examining the largest flour mill."[50] In fact, some Americans viewed it as positively undemocratic to exclude the public. One man chided a pair of English travelers for their reticence to sally forth to the Rochester mills without a letter of introduction. He bragged, "I recon [sic] there's no occasion for permission as we're too free in this country for that, I conclude you can go where you please."[51] Travelers who were able to identify themselves as persons of import, such as John Fowler, who was scouting settlements for emigrants, found local merchants and officials ready to guide them. A host's efforts could pay off handsomely. In 1838 the English industrial-urban reformer James Silk Buckingham was guided about Rochester by the mayor and the local historian. In Syracuse he was hosted by the director of the Syracuse Bank and the superintendent of the State Salt Works. Buckingham's accounts are filled with facts, anecdotes, and conclusions that came from booster sources. "The whole aspect of the place is that of one in which all that has been done, is well done," he reported of Rochester. "It will form an excellent nucleus for the accumulation around it of the materials of a great future city."[52]

With milling and manufacturing flourishing at the Middle Falls, a reinvigorated warehousing and shipping center developed in the original Rochester milling quadrant. The Erie Canal aqueduct, eighty rods south of the Buffalo Street bridge, routed the canal directly through the milling quadrant and looped around the commercial district before turning northward. Entrepreneurs constructed several turning basins along the canal's course, including a large one in the old milling district constructed by Jonathan Child (Fig. 9). Child also constructed four stone warehouses and five frame warehouses, which joined the remaining mills at the site. Merchant millers then developed an elevator system for transporting wheat out of the canal boats and into their mills. Beach's Red Mill, about one hundred feet from the basin, was fed wheat from a contraption that raised the wheat overhead and conveyed it horizontally to the mill.[53] Forwarding agents constructed new warehouses around these loading docks which, along with the remaining mills, reinforced the industrial cast of this district.

Yet the nuance in the reconfigured industrial district was already changing. The commercial aspects of milling became more pronounced with the bustle of forwarding agents conducting business in warehouses and offices, making the district as much a white-collar merchants' exchange as a blue-collar milling center. Facing the basin, the Rochester and Clinton House hotels and a new bank added to the functional and architectural tableau of the business (rather than the process) of industry (Fig. 13). Adaptable to changing economic priorities, in 1828 the local authorities renamed Mill Street "Exchange Street," recoding the district into a nexus of commerce and industry, in the spirit of a merchant's exchange. The city then gave the name "Mill Street" to a different street located by the industrially invigorated Middle Falls. (Being ever willing to adjust to changed circumstances, including reputations, the city renamed Carroll Street in 1831 to "State Street" following a rift between the city and the son of the recently deceased Charles Carroll.)[54] Exchange Street became the magnet for commission and forwarding houses, and its commercial climate was bolstered by the prominent hotels and their integral meeting rooms, newspaper offices, and nearby banks. This area soon became the dominant shipping and trading hub in the city. "If the Arcade was the forum of ancient Rochester," recalled resident Jenny Marsh Parker, "Child's Basin and slip was its Mediterranean, conditioning and vitalizing its canal commerce."[55]

Just as settlements congregated to the water advantage of the Genesee Falls, industries clustered around Onondaga Lake. The water advantage in this case was neither waterpower nor navigation but, rather, the salt springs that ringed the lake.[56] State-founded Salina, colloquially known as Salt Point, was soon joined by private speculation in salt-boiling, single-economy villages that ringed the lake. Syracuse investors, too, had not been able to resist the lure of salt money. By 1816 a salt-boiling establishment was in operation just two hundred yards from Bogardus Tavern. In 1822

Joshua Forman and Isaiah Townsend of the Syracuse Company traveled to Cape Cod to study the solar evaporation process of salt manufacture. The Syracuse Salt Company and Onondaga Salt Company shortly commenced salt making near the 1805 millpond, thus reinvigorating the waning trace the first industrial node.

One challenge was getting the raw material—the saltwater itself—into town. The solution was a mile-long log flume that carried brine from the salt springs ringing Onondaga Lake to distributing pump houses. Both the saltboiling and solar evaporation processes were carried out in structures quite distinct from the commercial, civic, or even other industrial districts (Fig. 15). Chimneyed salt blocks boiled the brine over wood-fed fires and produced fine salt. Open solar evaporation vats tapped the passive heat of the sun to create a coarser salt. One British visitor counted one hundred salt "factories" at Salina and twenty-three at Syracuse in 1831.[57] By 1834 Syracuse alone was producing nearly two million bushels of salt per year. By 1858 the salt production of Syracuse, Salina, Liverpool, and Geddes combined totaled over seven million bushels, most of it shipped out through the Syracuse transportation hub.[58] The feeder line of the Oswego Canal between Syracuse and Salina was heavily lined with salt blocks and their warehouses. Syracuse became known as "Salt City" not just because of its own production but also because it controlled the warehousing and shipping for the region.

The relative oddity of salt manufacture appealed to the city's tourists, whose accounts may have tempted industrial investors. Both Barber and Howe's 1841 gazetteer and French's 1860 gazetteer illustrated the various refining processes, assuming and highlighting public interest in the saltworks. Both artists included in their descriptive illustrations top-hatted gentlemen surveying the saltworks while gesturing to a visitor. The saltworks were encoded as more than a utilitarian landscape and transformed into a polite landscape of civic pride in local production and possible enticement to future investment.[59] Although Rochester's pounding falls and even its pounding mills could make for a sublime experience, Syracuse's tamer saltworks also demanded attention. Some travelers found themselves capable of waxing rhapsodic over the bushels of salt produced in Onondaga County. One English visitor dramatized the billows of salt boiling: "the immense volume of salt water is thrown up like a volcanic eruption on the immediate border of the beautiful fresh-water lake."[60] His compatriot Basil Hall found irony an easier reaction, dryly noting that "it was a very agreeable and novel sight to me to behold at this place upwards of 200 acres actually covered with vats filled with salt water in the act of evaporation."⁶¹ Either way salt was a place-making endeavor.⁶²

Reviewing the City

Rochester's 1825 celebration of the statewide opening of the Erie Canal displayed the sorted city's urban maturation while underscoring the stature of the merchants and lawyers who had guided the construction of a sorted cityscape. The combination of the physical staging and the selected participants acclaimed the particularly mercantile vision that had guided the economic and cultural colonization of the New York interior. Both the sorted city and the city leaders could be seen as taking center stage.

Despite the rain that October day, eager Rochester crowds lined the Erie Canal waiting for the festivities to begin. An official flotilla of twelve bedecked canal boats had left Buffalo the day before and was now carrying a host of luminaries and statesmen toward Albany and ultimately the New York City harbor. Canon salvos relayed the travelers' progress eastward along the canal. Outside Rochester eight uniformed militias announced their sighting by firing salutes. As the flotilla approached the Genesee aqueduct, sentries ceremoniously halted the procession, exchanged official greetings, and bid them to enter the city.⁶³

Ostensibly focused on the state's Erie Canal achievement, the Rochester Committee of Congratulation had also taken care to celebrate the urban achievements of their thirteen-year-old settlement by showing off their three distinguishable urban districts (Fig. 2). Local dignitaries greeted Governor DeWitt Clinton and Lieutenant Governor James Tallmadge at Child's Basin, which was lined with new warehouses and flour mills. Here they exchanged congratulatory speeches under a triumphal arch erected for the occasion. For prayers and a public address the assembled procession then marched one block west to the spacious Presbyterian church behind the Monroe County Courthouse (which was too small to host the occasion). For dinner and toasts they processed down Buffalo Street and up Carroll Street to Christopher's Mansion House hotel.⁶⁴ A few hours later the esteemed members of the flotilla reboarded their boats and, with Rochester's "Lion of the West" added to the procession, floated eastward. Parade paths and ceremonial locations are typically chosen because these places have symbolic meaning.⁶⁵ In this parade the committee routed the celebration through the milling and shipping district along the basin, the civic district of courthouse and churches, and ended at the hotel in the commercial district before looping back to the basin. These three close yet distinct districts showed that the upstart settlement of Rochester possessed the economic and urban maturity of a bona fide city.

The distinctly sorted cityscape was only part of the urban show. The day was also a carefully orchestrated social event designed to showcase the entrepreneurial and professional class that had molded Rochester into a commercial city. The white, male merchants, millers, and attorneys who had actively "improved" the hinterlands wrote themselves into the script. Two lawyers, Vincent Matthews of Rochester and John Spencer of Canandaigua, made the first official greetings to the governor at Child's Basin. Matthews's law office was located in one of the Greek Revival temples fronting the courthouse, and he was fully part of the boosterish professional class in Rochester. In contrast, Spencer, a one-time U.S. congressman, state legislator, and attorney general, necessarily glossed over the southern tier's initial opposition to the rise of Rochester, including the routing of state roads and the Erie Canal. "Altho' those whom we represent could anticipate no peculiar benefits from the work, which would rather tend to divert the sources of their prosperity," Spencer acknowledged, they had nonetheless given the canal their "cordial, vigorous, and useful support."[66] Transportation had knit the cities together, and the skein of mercantile networks ultimately boosted the region.

Another Rochester attorney, Timothy Childs, who delivered the public address at the Presbyterian church, painted the scene of a newly moralized as well as civilized landscape that accompanied its economic improvement. Because of the canal, "religion and literature are occupying every foot of soil." The canal forced the retreat of "unsubdued nature" and furthermore disciplined the "ignorant and idolatrous" behavior of "the intellectual being who is rioting upon all this bounty."[67] According to Childs, the canal lifted the pioneer settlement into higher standards of urban, personal, and spiritual cultivation. This was precisely the kind of Protestant affirmation that the merchant class was using to build its economic authority and social hegemony, and it resonated with the culture of improvement which conflated economic and moral progress.[68]

The twenty-seven toasts raised at the Mansion House hotel were part of a closed circle of commercial congratulations, given and received by "gentlemen" with economic interests in promoting the canal and their cities. Merchants and millers joined the lawyers in complimenting the canal and the men who championed it. State dignitaries returned the favor by praising the city that reaped its benefits. The urban fabric itself, after all, had been front and center in the organizers' plans and was graciously acknowledged by the toastmasters.

> *By his excellency De Witt Clinton.*—Rochester.—In 1810, I saw it without a house or an inhabitant.—In 1825, I see it the nucleus of an opulent and populous CITY, and the central point of numerous and transcendent blessings.
>
> *By Lt. Governor Tallmadge:* The village of Rochester: the *"Young Lion of the West."*—It stands proudly on a rock, here the most useful and noblest of streams laves its feet. Such in youth, its age promises to attain the acme of greatness.[69]

Only as the official flotilla was leaving Rochester did the festivities include public participation, as residents lit up the city. The local newspaper reported: "The illumination was general and brilliant. Transparencies, with patriotic inscriptions, were seen in several directions. Rockets were set off upon the departure of the boats, and there was a handsome display of fireworks." After the officials had departed, the celebration concluded with a Grand Canal Ball, "which was graced by all the fashion and beauty of the village and vicinity. Nothing occurred to mar the pleasures of the day. Everybody was both happy and grateful to enjoy the high honour of celebrating the completion of the GRAND ERIE CANAL."[70]

A more local celebration had occurred two years earlier, when the Rochester stretch of the canal was completed. Instead of focusing on the mercantilist potential of the chain of trade, the first orations of September 1823 celebrated the labor that had built the canal. A second celebration a month later was similarly local in its praise. A small flotilla of boats floated back and forth through Rochester to the Masonic tune "The Temple's Completed" and concluded with public prayers at Child's Basin. Although they were excluded from the banquet at Christopher's Mansion house, the women of the city were lauded in the final toast: "To the ladies of Rochester— the brightest ornament of our day's celebration."[71]

The 1825 grand celebration, in contrast to the earlier local focus, seized on the external, promotional opportunity for the city. The statewide event even had an official chronicler, the same William Leete Stone who had earlier made his own foray into the canal towns. City leaders knew there would be newspaper accounts of the festivities along the cross-state journey, and the Rochester Committee of Congratulation structured an event whose path would show the young settlement to its most sorted, that is urban, effect. The message was duly received and reported. Stone printed that

Rochester is "a rich and beautiful town, which disdaining, as it were, the intermediate grade of a village, has sprung from a hamlet to the full-grown size, wealth and importance of a city."[72]

In actuality, reviews of these new cities were mixed, both in their assessments and their honesty. Margaret and Basil Hall's sketch and writings captured both the truth and the lie of Rochester in 1827. Yes, it had a sorted landscape, but it was not entirely a polished place. The English couple were astonished to find "within the immediate limits of the inhabited town itself, in streets, too, where shops were opened, and all sorts of business actually going on, that we had to drive first on one side, and then on the other, to avoid the stumps of an oak, or a hemlock, or a pine-tree, staring us full in the face."[73] The city was a work in progress. What was built was often raw and poorly constructed. Even the proud citizen Edwin Scrantom admitted that early building in Rochester was "fast building, and shanty building of course."[74] The first generation of buildings was clearly constructed for the present, not for perpetuity. The overall effect of the growing and hasty city was not necessarily pleasing. "There is no great beauty about it," opined an anonymous traveler in Rochester during the 1830s, "and at this time I consider it a dirty place. All the streets are filled with mud and rubbish. Building is the order of the day."[75]

During this first stage of urban settlement residents and investors constructed the physical substance of the city and built up the initial spatial culture of the segmented city. Street, lot, building, and activity conjoined to define districts within the city. The first-stage districts had a look and a feel that identified the entire settlement, not just the district, as a potential if not quite realized city. The appearances of the districts were the product of conscious decisions by several different groups. The city builders, which included developers, investors, and settlers, constructed zones of functional differences. Certainly, the look of the new nineteenth-century city was not a coincidence. These districts were designed to reflect what the people thought made workable, and thus good, urban form.

This first stage of building the new nineteenth-century city was the emergence of the articulated districts, not their refinement. It was a landscape of mixed messages. Fresh and new, said one. Dirty and littered, retorted another. A few fancy buildings rubbed shoulders with the lowest sort. The districts were definable and recognizable but not consistent. The architectural and social ordering of that aborning cityscape would consume city leaders during the ensuing decades.

FOUR

Refining the Sorted City
Appearances in the Commercial District

By midcentury the new cities had firmly taken hold. In 1840 Rochester's population reached 20,200, and by 1858 it had doubled to nearly 44,000 residents. Syracuse was smaller but similarly prosperous. In 1840 Syracuse and Salina together had 11,000 residents, by 1858 over 25,000.[1] In 1841 John Barber and Henry Howe published the first edition of their illustrated gazetteer of New York State, which included "faithful representations" capturing the essence of each notable city (Figs. 16–17).[2] Both Rochester and Syracuse required several views, one of the city center and another of the mills and saltworks at the edge. By emphasizing the cities' commercial, industrial, and civic districts in their engravings, the gazetteers tapped into the cultural currency of the sorted cityscape as the urban standard. By emphasizing the distinct districts within the cities, Barber and Howe showed that thirty years after their founding the sorted cityscape continued to have cultural currency in the appraisal of a city. In fact, Barber and Howe specifically stated that Syracuse "now has a city-like appearance."[3]

The illustrations of the "central part" of both Syracuse and Rochester also showed that aesthetic standards had been raised in the commercial district. Showy pieces of architecture shared the street with uniform rows of buildings, evincing improved building standards and a heightened taste for beauty and style. The "faithful representations," of course, carried an equal dose of idealization in the extreme tidiness of the view, but the bias in the view making attested to a preference for that visual order.

Appearances had always mattered. The very premise of a legibly sorted city demanded that it be apparent, but the standards of what those appearances should be shifted after the first wave of settlement. Having survived the first urban shake-outs that saw some settlements disappear and others be absorbed into nearby rivals, settlers and investors set their sights on a higher standard that demanded architectural and urban refinements. City leaders largely rejected the notion of beauty for beauty's sake but pragmatically incorporated ideas about the utility of aesthetics into their building efforts.

The higher architectural standards emanated from an aesthetic shift that paralleled the rise of the middle class as the cultural authority in the cities. The merchant and professional class that had invested in the settling of the city assumed a sort of "merchant *oblige*" to improve it culturally as well. Entrepreneurship was in essence a sign of good citizenship in the new city's urban society, and the merchant's risk taking and economic investment led to his social capital. The intensity of entrepreneurial commitment, not necessarily the results, mattered. Even after a number of Rochester's leading merchants went bankrupt and were forced to vacate their costly homes and give up control of their businesses, they were not forced to relinquish their social status in the community they had led.[4] In aspiring to the mantle of urban tastemaker, the middle class worked at fashioning a cityscape that, not coincidentally, served their own twinned interests of mercantile development and cultural display. General urban boosterism continued to underlie much of their efforts, but it was a promotionalism that particularly targeted like-minded aspirants.

The attention on the three distinct districts which had percolated through the planning and initial building of the new city shifted to a tighter focus on the commercial district as a symbolic space during the second quarter of the nineteenth century. Reflecting its prominence within the traveler's orbit, the residents' daily lives, the merchants' world, and the very identification of the new settlements as commercial cities, the commercial district became the touchstone in urban evaluations. Based on the architectural and social scene at Clinton Square, one traveler declared Syracuse a "city-looking" place: "I was agreeably surprised to find this village (as the inhabitants with amazing modesty still continue to call it) in reality a very large and city-looking town, with wide and well-paved streets, lofty brick houses, fine stores, and an air of business-doing on an extensive scale."[5] Borrowing metropolitan metaphors, the commercial district symbolized the city. During the 1840s one traveler applied the Broadway analogy to Rochester. "Between sundown and nine in the evening, Main [Buffalo] Street, from State to the east side of the River, presents a fair and full miniature of Broadway in New York, by the throng of people passing to and fro. So dense is the crowd that one is compelled to elbow his way along the best way they can."[6] Vandewater's oft-reprinted *Tourist Guide* concluded in 1831 that Rochester was "the most extensive, populous, and important place in the western country. It has been termed the 'Western New-York.'"[7] The construction of new hotels and the continuous street wall of buildings in Syracuse similarly gave the city "the appearance of New-York in miniature."[8]

The commercial district was becoming the epitome of the city. The two terms were nearly synonymous. One Rochesterian recalled that the phrase "going down to the city" meant the same as "going to the Four Corners."[9] The creation in 1853 of a strictly business directory for Syracuse was evidence of not only the growth of the commercial district but also the district's importance to the overall stature of the city. Urban pride as much as commercial complexity, the publisher explained, compelled such a directory. "The features here delineated we think will render it a *vade mecum* with all classes of our citizens. Other cities have possessed such publications for years; and certainly it is high time to Syracusans to lay aside their *village* habits and ideas, and fill the measure of municipal importance."[10] The proof of a city lay in its business and commercial activity.

The merchant leaders targeted the commercial district for architectural and social refinements. Landmark buildings that responded to changing tastes in fashion became one vehicle through which the business community promoted the new settlements as not only economically vital but also culturally sophisticated, amenity-filled cities. Built on prominent lots within the commercial district and designed to be architecturally eye-catching, the hotels and largest stores became urban benchmarks for both residents and travelers. Stylish architecture helped move the reputations of the cities from simply being places of economic opportunity to becoming places of true urban and urbane life. Architectural design also codified the divisions between the adjacent commercial and industrial districts and reinforced the sorted template of the cityscape.

The aestheticized commercial district, however, was not just about architectural detailing. The practice of aesthetics in the new city included a spatial aesthetic as well. Tidiness and order were a "look" as well as a system that could be found in the repetition of a store's bays, the vertical plane of the shop-front walls, the uniform projection of shop-front awnings, and the unencumbered space of the pedestrian sidewalks in front. The transformation of the new city's commercial district between the first stage of settlement and the second stage of refinement showed how deeply an aesthetic of beauty and order dominated the merchants' architectural and spatial practices. The social implications of this refinement will be examined in the next chapter, but the focus here is on understanding the visual implications of the buildings as architectural set pieces and the roles such "sets" played in articulating the mercantile cityscape. Expanding the early-nineteenth-century creed that framed economic productivity as the key urban virtue, the second generation of city builders began to see beauty as an overlooked contributor to an urban commonweal.

Adding architectural beauty to the list of civic virtues was such a novel concept that its advocates struggled to find the right word to describe their holistic vision of what we would today call "urban design." The anonymous author of an 1830 article published in the *American Journal of Science and Arts* apologized for his inexact title of "Architecture," explaining: "my remarks will also take a wider range, and embrace a science, for which I cannot find a name, for the good reason, that among the nations from which we draw our language, no such science could be known. I mean the choice of position, and the planning of towns, with the grounds and appurtenances connected with them."[11] Although he was unsure about the label to put on his thinking, the author was quite confident in his subject, a critique of contemporary urban and architectural practices. He rejected the mercantile city, with its gridded streets and commercial priorities, and replaced it with his vision of a city built for visual effect and civic inclinations. Beauty, he charged, had been wrongfully neglected as an important consideration when laying out or building up cities. Although he named no particular offenders, Syracuse and Rochester were certainly vulnerable to his reproach.

The critic sketched out his antidote to the dullness of what he called cities of "squares or rectangular parallelograms," replacing them with his "*beau ideal* of a town" which combined convenience, symmetry, neatness, variety, and beauty (Fig. 18).[12] He specifically cautioned readers that his illustration was not a blueprint but, rather, a schematic diagram "to elucidate my remarks."[13] His proposed *beaux ideals* were far removed from the prosaic concerns that had guided mercantile city builders at the turn of the century. Instead of profits and efficiency, aesthetics and civic space dominated the diagram. He proposed a public green with a fountain, wooded alleés, a civic building on a central eminence, a street lined with churches and banks, a bluff capped by a handsome dwelling or a public monument, bending and offset streets for surprising views, and avenues focused upon distant pillars or obelisks. Some of these ideas echoed the monumental tradition of European grand manner planning, while others presaged late-nineteenth-century organic, townscape planning. Virtually all of them rejected the values and appearances of the mercantile city, which he described as "calculated for dull labor, or lynx-eyed gain."[14]

In the critic's urban vision the commercial was subordinated to the civic. His proposal was grounded in a clear valuing of shared civic spaces, available to all, and experienced through a meditative ramble or an inspirational vista. This genteel view was essentially an elitist one that omitted the role or space of industry and commerce in favor of nonproductive uses, but it was also a people-centered view that paid attention to the qualitative experience of place rather than the efficiency of it. Civic values, however presumptive, structured this educated and leisured commonwealth of bourgeois citizens who had the time and sensitivity to appreciate the artful plans.

Although the ideology of such a city was antithetical to the mercantile city that subordinated the civic to the commercial, the argument for aesthetics did not go unheeded by the second generation of city builders, who found ways to embellish their commercially conditioned public space. The first generation of settlers in Syracuse and Rochester had not incorporated the obvious markers of civic space, such as parks, civic squares, or monuments, which the anonymous critic argued would inspire people to love the city. The second generation showed no inclination to abandon its "parallelograms" of commercially productive broad streets for the curving sway of a charming boulevard. Yet an acceptance of the pragmatic utility of appearances ran deeply through both generations. The deliberate construction of a sorted cityscape acknowledged that appearances mattered, but it was no longer enough just to show that an aborning city boasted civic, commercial, and industrial districts. The cityscape now needed to present a certain kind of sophistication which showed that the cities were economically and culturally au courant and not a stagnating or primitive backwater.

For a new city still competing for human and economic capital, architecture was a tool of enticement. Mobile Americans had a choice of where to live and where to stake their future, and they sized up a community in part on its buildings. As the anonymous urban critic had warned: "We are a calculating people, sufficiently attentive to present interests. . . . We are also a travelling people . . . with abundant opportunity for comparing places and scenes with one another, [and] we are gradually forming a pretty correct judgment, as to the beauties of a landscape or town. . . . This is shown in the crowds that gather to the deck of a . . . canal boat, as a . . . handsome village is approached and is heard in the murmur of approbation among little groups of such travellers."[15] The peripatetic public was a discerning one, inclined to pass judgment on the newest of settlements. Thus, "before many years, he who will wish a town *to flourish* . . . will have to consult not only health and convenience, but also beauty and good taste."[16] Settlements courting a population needed to cater to a public with an ever increasing urban and architectural literacy.

The owner of the Erie Canal "Rocket Lines" invoked aesthetic pragmatism in his request to the New York canal board for permission to build a ticket office on the state-owned berm of the canal embankment. Proposing a one-story building of cut stone ornamented by an iron railing, Van Patten explained that the Syracuse city authorities had indicated "if it were under their control, they should cheerfully grant the Petition as they thought an office of that description would not only be useful but ornamental."[17] Attention to aesthetics, particularly those held by the scouting public, was of instrumental value to building a city.[18]

Even the landscaping of the cityscape became a cause for improvement. With the first days of swamp and wood now behind them, residents adopted a more forgiving attitude toward nature. The first generation cut down trees to clear a city, and the second generation planted them to beautify it. Without a trace of irony, in 1829 a Syracuse newspaper urged its citizens to ornament the urban landscape by planting trees. "There's nothing which contributes more to the embellishment and comfort of a town than the growth of shade trees. . . . Every additional tree tends to beautify and improve the city and to increase the value of individual property."[19] Beauty had pragmatic utility. The call for a landscape within the cityscape was issued again in the late 1830s by E. W. Leavenworth, the president of the Syracuse city council. Playing the expected role of a spirited public servant, he was "always zealous in the interest of the aesthetic side of improvements and labored for broad streets, more parks, and shade trees."[20] Under his leadership the elite residential squares of Fayette Park and Forman Square were protected from the incursion of turnpike and railroad, and the railroad company was required to plant shade trees along its path through Washington Street.[21]

The author of "Architecture" was not the only critic calling for a new focus on architecture as both a symbol and method of achieving moral culture. In 1841 the nationally disseminated, Rochester-published *Genesee Farmer* urged families to pay attention to the message that their homes communicated. Viewing the derelict state of housing along the canal, one author chided, "here was no decoration, and I argue concerning this settlement, that there are no intellectual pleasures, no taste, no gentleness, no fireside happiness." The author assumed that "every reader has many times seen the same thing, and some have already learned the connection between simple decoration and domestic virtue and peace."[22] The same theme of the moral power of design was picked up in Catherine Beecher's treatise on domestic design, in which she stressed the need for health, comfort, and beauty in the home. "For while the aesthetic element must be subordinate . . . it yet holds a place of great significance among the influences which make home happy and attractive, which give it a constant and wholesome power over the young, and contributes much to the education of the entire household in refinement, intellectual development, and moral sensibility."[23]

If domestic architecture could be linked to domestic virtue, then it followed that public architecture measured public virtue. Indeed, following her short stopover in Syracuse, the author Eliza Steele concluded that, although the city was "still in its tens," it nonetheless presented "several good churches, a pretty court house, substantial warehouses, numerous shops and dwellings, with a lyceum and high school, so that it would seem the inhabitants ought to be wealthy, refined, and well educated."[24] Architecture was a window into social, cultural, and moral character of the settler and settlement. Without losing the core principal of economic growth, urban planning principals in the new cities were adding the aspiration of a cultural morality that conflated character and style.

Once again the business class took charge of the physical buildup of the city's public face. The new cities had started out as explicitly mercantile ventures, and the merchants' interests had been equated with the public interest. As the cities and popular ideas of good urban design evolved, the commercial leaders continued in their leadership role, this time adopting the aesthetics of high style and tidy order in their architectural practices. In 1840 David Buel, an attorney and businessman in Troy, New York, called for his city's architectural rejuvenation. Troy's second-generation residents were in a similar position to review their course of urban progress since the city's 1797 founding. Buel urged his fellow businessmen to meet their paternalist obligation of urban improvement by becoming patrons of the arts and commissioning institutions and edifices whose purpose and style would add honor to the city:

> In modern cities, the highest achievements in the useful and ornamental arts have been made under commercial patronage. Even in the middle ages, the finest specimens of architecture, the best models of paintings and sculpture, and the most splendid collections of books, were called into existence by the commercial communities of Venice, Bilboa, Naples and Florence. At this moment both in Europe and our own country, the architect, the painter, the sculptor and the author, must seek his principal patronage in commercial communities. It is time that our city should take some decisive steps in rearing institutions calculated to improve the intellect, cultivate the taste,

and afford occupation and enjoyment to the minds of those who now inhabit the city, and of those who shall come after us. The beauty and elegance of our streets and dwellings, excite the attention of strangers.[25]

The notion that the business community should lead in the patronage of the arts, coupled with the prominence of the commercial district as the most frequented and noticeable district in the city and the steady infusion of capital into commercial affairs, all translated into businesses being the first to receive architectural improvement. In a piece of architectural criticism published in 1864, James Jackson Jarves reviewed recent building trends and lauded the rise of "solid and handsome blocks of stores, in more or less good taste, appropriate to their purpose, effective as street architecture, and novel in many of their features."[26] Although personally disappointed by the crass emphasis on commerce, he was encouraged by any signs of architectural affection. "If a knowledge of the fundamental principles of architecture equaling the zeal displayed in building could be spread among all classes, a better order of things would soon appear."[27] Commercial buildings had acquired the patina of honorable exemplar, and the paternalist merchant was now a good citizen who constructed credits to his community. Such merchant *oblige* not only served the cause of the commercial city but also reinforced the merchant's cultural leadership within the social hierarchy of the city.

Hotels and the Commercial District

An inn was part of the grander mercantilist scheme of improving the interior. Inns were the first buildings constructed as intentional landmark pieces of architecture intended to promote the city to outsiders. Hotels provided necessary services for residents and served as stage sets crafted for bourgeois sociability. First-generation inns were harbingers of, even prerequisites to, urbanization. A backwater inn, explained the historian Daniel Boorstin, was designed "not to serve cities but to build them."[28] Catering to both settlers and travelers seeking room and board, inns were necessities in the sparsely inhabited regions. Inns were recognized as critical settlement building stimuli, so much so that the 1804 contract between New York State and Abraham Walton stipulated that he must erect an inn on his 250-acre tract. However rustic, Bogardus Inn served not only the new settlement that would become Syracuse but also all the towns in the vicinity of the Seneca Turnpike. The state's belief that an inn would stimulate development was repeated in new towns across the country, leading the Englishman Anthony Trollope to note: "when the new hotel rises up in the wilderness, it is presumed that people will come here with the express object of inhabiting it. The hotel itself will create a population." In the United States, he concluded, towns "run to the . . . hotels."[29]

A town's first inn could be quite rudimentary. William Leete Stone suffered his first night in Syracuse in that state-mandated inn built by Bogardus.[30] Despite the close quarters on the canal boats, some travelers, such as the emigrant scout John Howison, preferred to spend the night there rather than to endure Bogardus's old inn, hardly good press for the new settlement.[31] The inn may have served the immediate local needs but lagged in its ability to promote the settlement to newcomers.

Within a decade of settlement innkeepers began to invest in not only increasingly stylish and accommodating quarters but also, quite simply, more hotels. A cosmopolitan Argentine taking the Erie Canal passage concluded that any town of substance had at least two hotels.[32] Nathaniel Hawthorne was more amused than persuaded by the quarters he saw on his 1835 Erie Canal voyage, noting that every urban aspiring village held "generally two taverns, bearing over their piazzas the pompous titles of 'hotel,' 'exchange,' 'tontine,' or 'coffee-house.'"[33] Indeed, both Syracuse and Rochester quickly displayed several hotels, including showy hotels in prominent locations and farmers' inns on the edges of town.

The conduct of business, the running of local government, and simple sociability all happened in the downtown hotels. It was almost as a secondary feature that hotels rented out lodgings and served meals to the overnight travelers and long-term residents who boarded rather than setting up household. The hotels' intensive use day and night gave them a special prominence and visibility in the city. As such, city promoters targeted them as the first buildings to be constructed and remodeled. These second-generation hotels, with their stylish facades and increased public amenities of public halls and parlors in addition to barrooms, pushed a settlement one step closer to architectural and institutional urbanity.[34]

Hotels were integrated into the conduct of daily business. In a city bereft of other public space, the bars and parlors were typically the primary meeting places, both official and spontaneous. The public sought out the ground-floor common room, with its assortment of newspapers, fliers, residents, and travelers who might offer up promising bits of news, gossip, and insights. When the Rochester proprietor of the United States Hotel on Buffalo Street remodeled his facility at the outer edge of the commercial district, he said little about the sleeping chambers but reminded the public that it was "a convenient resort for men of business, as it is but a few minutes walk from it to the Banks and Post-Office" that were located at the Four Corners.[35] Similarly, shopkeepers used the Syracuse House and Mansion House hotels as coordinates for locating their businesses in town.[36] The most prominent hotels rented quarters to banks. Located at the Four Corners, the Rochester Eagle Hotel featured the Bank of Monroe in its corner office (Fig. 31). Prominently linking Clinton and Hanover squares, the corner Syracuse House rented quarters to the Onondaga County Bank.

In the absence of city halls, hotel assembly rooms were deputized for the running of local government business, prompting entrepreneurs to construct "courthouse" hotels to tap into this market. After the Monroe County courthouse was completed in 1822, the Monroe (later National) House hotel arose across the street. Following the construction of the compromise courthouse in between the two settlements of Salina and Syracuse, the Center House hotel (aptly named for being centered between the two cities) arose next door. The sequence was reversed in 1856, when Colonel Vorhees maneuvered to get the new Onondaga County courthouse constructed next door to his hotel at Clinton Square. Particular hotels became aligned with specific political parties. During the 1820s the competing Clinton House and Rochester House hotels both opened on Exchange Street facing Child's Basin. Their names were honorific, paying respects to the city founder, Nathaniel Rochester, and to the Erie Canal champion Governor DeWitt Clinton. But they were also factional, alluding to the pro–Van Buren Republicanism that Nathaniel Rochester backed, in opposition to the Clinton-associated Federalism favored by other prominent merchant-millers in the city.

In 1820, when a new hotel was planned diagonally across Salina Street from the aged Mansion House, its builders were aiming for a quality piece of architecture on a prominent corner which promoted the economic potential and cultural cachet of Syracuse to outsiders. The decision to build solidly and stylishly was based in part on the investors' conviction that the urban experiment was going to succeed and thus a more permanent building was warranted. It was also based on the boosterish conviction that architecture itself could promote more settlement and a better caliber of settlement. The propagandistic aspect to the new Syracuse House was acknowledged from the start. As one of the builders recalled, when the partners approached Joshua Forman to purchase the prominent lot at the hinge between Clinton and Hanover squares, they all agreed that something remarkable should be erected: "Judge Forman was anxious that we should put up the best hotel west of Albany, as he thought it would be an inducement to others to purchase lots and start a village."[37] A sophisticated piece of architecture at the juncture of Clinton and Hanover squares, at the juncture of the canal and the turnpike, was perfectly placed to grab public attention. By the 1830s an estimated one thousand people a day passed through Syracuse by stage or canal.[38] Hotels physically and impressionistically mediated a traveler's experience of a place. The better the initial introduction to the city, the reasoning went, the better the ultimate verdict.[39]

Completed in 1822, the brick, two-story, fifty-by-fifty-foot hotel eclipsed the frame inn built a generation earlier by Bogardus. An engraving drawn from the builder's recollections showed a simple Federal-style building aggrandized by its brick construction and the contrast with its surroundings (Fig. 19). The interests of the hotel were inextricably tied to the interests of the city, and it was no coincidence that the name they chose for the new hotel was "Syracuse." The Syracuse House stood poised to intercept travelers from the canal or turnpike and to persuade them of the city's opportunities and civility.

It was a rocky start. Travelers were not impressed. One guest complained about the discrepancy between exterior polish of the "large building . . . promising in its appearance" and the disappointing accommodations: "oh such beds they had probably not been changed for the last month and were plentifully provided with every etcetera to render them uncomfortable."[40] A knickerbocker passing through breezily dismissed Syracuse as "rather on the decline if we were to judge of the appearance. . . . Hotel in a ruinous condition, a large substantial brick building, but scarcely a window in the establishment, which had not glass broken. Panes of glass mended by stuffing old rags in them. Economical in the extreme."[41] One of the builders died in a construction accident, and by 1827 his partner was forced to sell out to the Syracuse Company, the group of Albany investors who in 1824 had purchased the many unsold Walton Tract lots. Far from succeeding as an urban beacon, the Syracuse House bespoke of urban difficulties.

The better-capitalized Syracuse Company shared the convictions of its predecessors in the power of architecture as a place-making device. William James of the Syracuse Company declared that "the house was too low; that he would take it down and put up the best house in the State."[42] Resident Amos Granger urged them on in their building activities. "I am glad to hear that you have concluded to improve the sheds and back yard to the Syracuse House, both your interests and Rusts [the innkeeper] require this improvement."⁴³ As a nearby lot holder, Granger was personally interested in finding ways to stimulate the area; he had already been rebuffed in his attempts to get the Onondaga County courthouse located in Syracuse and was relieved to find an active champion in the Syracuse Company.

The rebuilt hotel was dramatically different from its predecessor and neighbors (Figs. 20–21). The new brick building rose four stories with a street-level veranda and continuous projecting galleries running along the facade at each story. The addition of three-story wings on Salina and Genesee streets further extended the hotel's urban presence and services. The showy Syracuse House was an architectural showpiece for a generation and a familiar landmark for a century. A picture of it even graced the locally minted three-dollar bill. One resident was dumbfounded: "It was the wonder of all that such a fine building should be put up in such a place. It was like a bouquet in a mud-hole."⁴⁴

The flowering of such a bouquet indicated an agenda beyond simple utility. There was little local demand for such outsized and outlandish accommodations. When the Syracuse Company rebuilt the Syracuse House into a pretentious structure, it sacrificed short-term losses for long-term gain. The hotel was a showpiece to inspire local boosters and to bait travelers and travel writers into seeing Syracuse as a settled, mature, sophisticated, and amenity-offering place. This time the interior was as polished as the exterior, including a dining room whose walls and ceiling were "tastefully ornamented in continuous war scenes, which were pleasing and attractive."⁴⁵ The promoters were savvy. The hotel won plaudits from travelers and helped put Syracuse on the travelers' map. Vandewater's travel guide specifically praised the new Syracuse House as one of the best pieces of architecture in western New York. And it was following a satisfying tea in the Syracuse House that Eliza Steele ventured her favorable opinions about the caliber of the architecture and people of the city.⁴⁶

Located just south of the canal, the Syracuse House became the pivot between the original commercial district in Clinton Square on the west side of Salina Street and the expansion of commerce into the adjacent Hanover Square on the east side of Salina Street. It helped open Hanover Square for more intensive investment at the same time it flamboyantly anchored the southeast corner of Clinton Square. By giving luster to both zones, it doubly promoted the commercial potential of both sites as well as Syracuse in general.

As the primary property holder in Syracuse, the Syracuse Company had the interest and financial ability to influence the physical and architectural development of the city. With the Syracuse House, they led by example, just as the architectural critics David Buel and J. J. Jarves had hoped. By setting an example, other lot holders might also be inspired to upgrade their buildings. One of the first ripples was the remodeling of the rival Mansion House, "a shabby patched up old concern" left over from Bogardus's pioneer days, which fared poorly in comparison with the new hotel.[47] Substantially remodeled by the new innkeeper, the Mansion House also offered amenities including a large assembly hall. According to Vandewater's guide, a traveler's first choice should be the Syracuse House, "a very extensive well-furnished hotel, and kept in the first style," but the refurbished Mansion House was a respectable second.[48] In 1845 the Mansion House was razed and replaced two years later by Vorhee's Empire House hotel and its attached business block (Fig. 24).

The cascading benefits of a showy hotel were similarly evident in Rochester. In 1817 the Scrantom's hewn log cabin on lot No. 1 was sold to make way for the Ensworth House. Although it was only frame, it made a large impression on visitors. A French traveler duly published his amazement at the cultural refinement found in a one-year-old hotel situated in a six-year-old settlement: "[I sat] at a table as delicately as it was correctly served. As in England, the forks are of steel and the spoons of silver. After dinner the cloth is removed and the table of well polished mahogany is covered with dessert which ordinarily consists of excellent native cheese, more or less ripe fruit, and berry preserves. It is at this moment that the conversation becomes animated, inspired by the Madeira wine which circulates around the table in crystal flagons. Who would not be astonished at so much luxury and refinement in a city which boasts but a few years existence?"[49] By the late 1820s, in the face of rising competition by the flurry of hotel building in the commercial district, the Ensworth House needed to be rebuilt. The brick, patriotically renamed Eagle Tavern was not its last improvement. The tavern underwent yet another gentrification during the 1830s and 1840s, including a name upgrade to the Eagle Hotel, classical porticoes on the exterior, and refurbished parlors on the inside (Figs. 31–32). By then nine business hotels clustered near the spine of Buffalo Street, and all this hotel building prompted the *Albany Journal* to report that Rochester's hotels "would reflect credit on any city."[50] Indeed, that was one of the key points behind the whole rebuilding: to promote the city to the outside world. Although Edwin Scrantom would publish in 1843 a piece of sheet music, "My Early Home," with his family's first Rochester cabin depicted on the cover, the loss of the ancestral home was but a sentimental trope (Fig. 30). The log cabin so picturesquely drawn had charm precisely because it was long gone. Replaced by the Ensworth House and then the brick Eagle Hotel, the change in construction on the prominent Four Corners lot was the best sign of urban progress.

Commercial Rows and Business Blocks

Syracuse had rough beginnings, acknowledged one resident: "the place made no progress until the Syracuse Company built the Syracuse House."[51] As hoped, the architectural improvements initiated with the hotels spread to other buildings in the commercial district. In 1838 John Townsend of the Syracuse Company reported his satisfaction with the city's improvement. Writing to his local agent, Townsend opined: "I have lately returned from Syracuse, where I have been attending to the business of the concern. The general appearance of the village is much improved & I think things look promising."[52]

Architectural appearances had always mattered to the Syracuse Company. After rebuilding the Syracuse House hotel, it continued to champion stylish buildings to bolster the image of the central city. Following an 1834 fire in Hanover Square, it worked with the owners to rebuild the low wood-framed shops into four-story brick buildings known as the Phoenix Buildings (Fig. 23).[53] Although the south side of Hanover Square had not been damaged in the fire, the owners of these buildings also upgraded their holdings to keep pace with the Phoenix Buildings. Uniform rows of brick buildings three and four stories high with similar architectural detailings transformed the architecturally insignificant south side of the square into the titled Franklin Buildings.[54] The Syracuse Company engaged in rebuilding its own properties as well. In 1842 it razed its range of wooden, one- and two-story buildings on the south side of Clinton Square and erected the Townsend Block, a three-story Onondaga limestone edifice a full twenty bays wide featuring a central pediment that vaguely alluded to classical edifices and thereby nudged commercial architecture toward a more monumental, landmark-caliber building.

These changes served the new mercantile cities in three important ways. First, the construction of business rows and blocks underscored the ongoing significance of a sorted cityscape to the investors who were building the city. As the Barber and Howe illustrated gazetteer depicted and as

Vandewater's travel guide concluded, these business rows clarified the first generation's outline of the business district and gave it a distinctly commercial appearance, "the appearance of New-York in miniature."[55]

Second, the larger size of the buildings both abetted and expressed the expanding scale and capitalization of business in the city. In just five years, between 1842 and 1847, six sizable business blocks were erected in Syracuse, and the rapid pace of building continued into the 1850s and 1860s. The small "shop house" had given way to the "business row" of uniform shop fronts, typically only three bays wide but forming a contiguous row with its neighbors. Each business row typically housed several tenants within each slice of the building. The pumped-up "business blocks" were a midcentury variation on the business row. They were larger and visually conceived of as a unified whole, but internally they could continue the practice of quartering a variety of businesses or introduce the practice of a vertically integrated manufacturing, wholesaling, and retailing operation.

Third, the architectural style of the buildings was the merchants' response to the new calls for architectural beauty as a civic good, particularly in commercial matters. The accentuated style of the new buildings, coupled with the repetition of buildings and the regulation of frontages and sidewalks, reflected a heightened aestheticism that pragmatically catered to multiple constituencies who cared about appearances and taste. The art critic's encouragement, the merchant's pride, and the traveler's critique, converged on the shared values of beauty, regularity, and order.

During this period of architectural refinements in the new cities, from approximately the 1830s to the 1850s, merchants consciously introduced national trends in building style and type. Building first in a plain, flattened, Greek Revival style and then in the more ornamental, molded qualities of the Italianate, merchants constructed buildings that defied any urban identification as either upstart or provincial. The matching brick rows on Hanover Square were part of a national trend in urban architecture (Figs. 23, 24). The business row building type with its plain, workmanlike architectural details was repeated across Syracuse, Rochester, the state, and indeed urban America. These 1830s commercial buildings broke from the 1820s shorter, narrower, gable-roofed, shop houses that had abutted their neighbors without much regard for the aggregate view. The new commercial idiom was typically a three-bay, three- to four-story brick structure with solid piers at the ground level enframing French doors or large display windows composed of small panes of glass. The upper stories were pierced with regularly spaced sash windows set within flat stone sills and lintels.

The cornice of the shed roof was modestly embellished with brick dentils. The pattern of building was not just uniform; it was uniformly simple. The minimalist Greek Revival business rows rejected the ornate columned porticoes that graced the high-style merchant dwellings, but the bold simplicity of the rows was itself a confident statement of Spartan restraint and financial prudence. The workmanlike buildings outlining Clinton and Hanover Squares projected an image of a solid, flourishing commercial center and, by extension, city. The uniformity of the new business rows belied the fragmented pattern of landownership of the individual lots and indicated the consensual, or at least normative, vision of the way a proper commercial row should appear.[56]

Fashion is notoriously fickle. Writing later in the century, when architectural fashion mandated more ornamentation, the 1820s settler M. C. Hand felt that the Phoenix Buildings had not aged well. He described them as being designed by architects with "little conception of beauty or elegance."[57] Styles might change, but the power of style had not. By the 1850s speculators turned to the latest Italianate style, with its larger plate glass windows, and boldly molded sills, lintels, and scrolled bracketed cornices. New buildings in the newest style not only replaced older buildings in the city center but also were used to promote real estate ventures pushing at the margins of the brimming commercial districts, such as the new Pike Block on the corner of Fayette and South Salina streets (Fig. 29).

After purchasing a double lot in the middle of the block of South Salina Street between Fayette and Jefferson streets—three blocks south of Clinton Square—the entrepreneurial Henry Dillaye embarked on a study tour in New York City and Philadelphia looking for ideas to attract businesses and customers so far south of the canal. Dillaye returned with the plan of unifying the entire length of the block with uniform Italianate facades. Leading by example, he personally developed the two middle lots, constructing a five-story building trimmed with the latest in fashion, including four plate glass windows twelve feet high, molded cast-iron window hoods and sills that enframed the upper-story elongated windows, and scrolled brackets embellishing a projecting cornice. Dillaye's improvements did not stop at the stoop; he removed the brick sidewalk and installed large flagstones in front of his store. Using his edifice as a prototype, Dillaye contracted with six other entrepreneurs to build like-designed buildings on either side; the sum of the parts formed the "Washington Block" (Fig. 28). The grand business block not only stimulated the economic development of South Salina Street, but it also reflected the new city's architectural literacy and, hence, economic and cultural parity with the established cities. As befit a city in the mercantile chain, this city's architecture reflected the movement of not only goods but also ideas.

Ambitiously designed business blocks likewise arose in Rochester as merchants sought to maximize their potential through aggressively showy commercial architecture. The local press heralded merchants for their gifts to the street. In 1848 the newspaper praised businessmen for the rarity of hiring an architect to construct their new brick, three-story block on State Street. Best of all, the newspaper stated, "we have guarantee, in the liberality of the proprietors, and the taste and the skills of the architect, for the belief that the new buildings will be ornaments to that section" of the city.[58] Improvements in commercial architecture were eagerly followed. Situated at the corner of Main and Liberty streets, on the east side of Rochester, the 1850 Emporium Buildings garnered much attention. Comprising a four-story brick building that was divided into three separate storefronts of unequal width, each store had its own distinct plate glass windows, including a rare circular one at the corner. Stone lintels hooded the windows, and the facade was enlivened—"filled up" was the description—with cast-iron ornaments.[59]

Looking back on Rochester's transformation, the pioneer settler Edward Scrantom took pride in the improvements to his city's commercial architecture. The process began with facades that had started out with "small, low doors, surmounted with fan-lights" only to be transformed into "larger doors with shutters, and the 'arcade' doors, which was the last style." Improvements in iron and glass advanced a more open aesthetic of smaller structural members enframing larger sheets of glass. Storefronts were transformed by "the removal of the old fronts, with their heavy piers, and substituting in their places the small fluted iron columns. The small glass windows gave way for the larger glass—then the bow window and the square projecting one with its four large lights, or it may be only one light." Building upkeep joined with salesmanship. "Then the outside embellishing; the paint and the putty, the iron awning frames, the gratings over the sidewalk openings, and the plant balustrades on the roof, to represent another story and afford advertising signs in great letters, to be read at a distance." The upgrading migrated to "inside improvements" as well with "the removing of old partitions and putting in iron columns for supports, and back additions for sky-lights, and frescoing, and papering, and painting." Architectural beautification worked hand in hand with business promotion. "These have been going on for the last twenty-five years with great profit to builders, mechanics, and artificers of many kinds, and with renewed satisfaction year by year, to new adventurers in trade, anxious to put the best foot forward and keep up with the times."⁶⁰ Presentation, not just presence, now mattered.

Double Fronts

Within the compactness of the sorted city, the divisions between districts could be even more narrow than the slight alley that separated the Monroe County courthouse from the Buffalo Street shops. In the cities in which shipping formed a major part of the economy, property owners constructed single buildings oriented to two different economies and transportation paths. The construction of double-fronted buildings—one facing the commercial-oriented street and the other facing the shipping-oriented canal or river—showed just how narrowly the dividing lines could be drawn in the sorted city.

Bifurcated buildings reflected both the commingling of retailing and wholesaling practices in a single site as well as the desire and ability to segment those differences. In the newest settlements most store owners were primarily wholesalers who supplied the needs of the surrounding settlers, who were themselves trying to carve out a living. Some of a storekeeper's stocks would, of course, have been sold at retail to local end users, but wholesaling protected the merchant from fluctuations in the local economy.⁶¹ Wholesalers who were selling from a warehouse needed to be close to roads or the canals for efficient importing and exporting of goods. Retailers needed to be conveniently clustered for the foot trade. In addition, buildings straddling the commercial and industrial districts were conveniently situated for the practice of a merchants' exchange.

Syracuse's Clinton and Hanover squares had many double-fronted buildings that straddled both the waterfront zone of warehousing and shipping and the street zone of pedestrian and wagon trade. After the 1819 replatting of Syracuse in response to the coming canal, the layout focused on the Erie Canal as its "Main Street," and buildings were designed to maximize both canal and street opportunities. Towpaths edging the canal provided a narrow buffer between building and water, and in cases of a canal turnout a wide, planked towpath doubled as dock and sidewalk. Lots stretching between canal and street presented the multiple advantage of tapping into varied vehicular access and thus could be divided into halves, with retailing along the street and warehousing, wholesaling, and shipping along the canal.

Central city proprietors constructed double-fronted buildings that capitalized on and delineated the distinctions between the two districts (Figs. 23, 25–27). Street facades featured large open fronts of display windows separated by stone or cast-iron piers, with evenly sized and spaced windows on the upper stories, an occasional hoist in the roof, and projecting cornices. Canal facades that fronted planked towpaths often followed a similar pattern of open fronts framed by sturdy piers, but, rather than being updated with larger, showier, and more vulnerable large plate glass, they presented plainer shop fronts of small-paned windows. At first glance the fenestration of the upper stories of the canal-fronted buildings replicated that of the retail side, but in place of the common three-bay window pattern was a staggered pattern of windows flanking an enlarged loading bay. Chutes often projected from these upper-story openings, and hoists and pulleys were built into the cornices. The warehouse stories were less heavily signed than retail fronts as well.[62]

Standing on the end of a row in Syracuse, the Journal Building actually presented three fronts, each expressively designated according to its function and audience. Hierarchies of finish expressed its different functions, audiences, and districts (Fig. 27). The Clinton Square facade featured a number of ornamental embellishments: a raised basement, pilasters, a corbeled cornice over a blind arcade, arched windows on the third story, and paired windows in the center bay. A separate staircase rose to the second-story printing offices, located above the higher-rent retail space at street level. It was, however, divided down the middle, reflecting the way in which this elevation acted as a hinge between its Water Street retail shops to the south and its Erie Canal warehouse to the north. Decorative features on the Water Street side of the facade were omitted on the canal side: the stringcourses were omitted, and the quoin-like pilaster at the corner did not wrap around to the canal side. The Water Street facade (unfortunately not visible in Fig. 27) probably featured the street-level pier and window pattern found on other retail buildings of the period. The Erie Canal side was simple, although being directly on Clinton Square made this facade part of the noticeable landscape, and so it received more detailing than canal-side fronts farther down the canal. A corbeled cornice wrapped around the street and canal sides, and the pattern of arched third-story windows was carried around as well. But neither pilasters nor expensive plate glass windows embellished the water side. Instead, signs of warehousing indicated the loading and unloading of goods. Hoists hung from the adjacent structure; barred windows and solid doors secured the contents. In fact, the ground floor was used as the local police station in the early 1850s.[63] Retail stores, a printing shop, a newspaper office, and a jail made for an unusual building type but one whose different functions were nonetheless architecturally differentiated. More important, the mixed-use Journal Building points to the variety of activities and people in the commercial center and the attempts to order them rationally through architecture.

The Appearances of Order

The prime aesthetic transformation of the cityscape during the second generation of city building was not simply the adoption of formal styles, although certainly Greek Revival and Italianate details cloaked the downtown. The widespread adoption of these styles was just one expression of a deeper aesthetic that valued uniformity over variety and order over license. Tidiness became a positive visual attribute. One of the greatest praises for the rebuilding of Hanover Square, for example, came from a resident who viewed the 1834 fire as a favor, "as it enabled the owners to rebuild more substantially and in greater uniformity."[64] Certainly, the Phoenix and Franklin buildings had much in common, with their brick construction, three- and four-story height, flush facades opened up by stone piers, and rhythmic fenestration with simple sills and lintels.

A taste for uniformity was evident in urban planning as well. In the debates about rebuilding the Salina Street bridge over the canal in 1833, a group of residents decried one proposal to widen it only toward the west. "Any departure from the centre will greatly mar the beauty and regularity of such street, as it will not be in line with the other bridge . . . and will in other respects have an unnatural and unpleasing appearance."[65] An 1851 Syracuse ordinance specifically stipulated that "it was in the permanent interests of [the] city" that newly added streets and squares must conform to the existing pattern both for public convenience and so that "uniformity may be produced."[66] Visual uniformity was not a by-product; it was a goal. It evinced a new appreciation by the city leadership for a more tidy and legible cityscape, one that preserved the commercial ethos of the mercantile city while nodding to the civic values of ordered public space. In the commercial city order and uniformity were equated with beauty.

A comparison of two views of Rochester's Buffalo Street, one made by Basil Hall in 1827 and the other by Barber and Howe in 1841, charts the sorted city's evolution from raw and snaggletoothed to refined and orderly (Figs. 8 and 16). The functional sorting is the same, but the architectural refinement of the tableau is marked in the second-generation streetscape. Hall's 1827 sketch focused on the contrasts between the commercial and civic districts. It captured the rather lackadaisical and indeterminate character of the commercial buildings, whose builders used close spacing and signs rather than building type or style to express their commercial purpose. No real sidewalks are evident, and the street itself is uneven. Even with the showy courthouse, the scene was ambivalent, both boasting the impressive urban achievements of only fifteen years and confessing that there remained much more to be done.

Barber and Howe's 1841 engraving included the commercial and civic elision, but this time there was nothing about the view to suggest a disorderly pioneer settlement or tenuous commercial district. Looking eastward, Buffalo Street cut a broad, level swath through the center of the view. Nathaniel Rochester's planned commercial and civic districts were still in place, although merchant storehouses had leapfrogged west of the courthouse and pressed eastward over the Buffalo Street bridge. The ramshackle commercial row in Hall's view had been replaced by the monolithic Smith Block. A small grove of trees sandwiched between two temple-fronted offices buffered the courthouse from the busy street. Two-story, two-bay, gable-roofed frame buildings persisted, but brick, four-story, multiple-bayed, shed-roofed buildings dominated the streetscape. The new buildings initiated an urban wall of contiguous, same-height facades. The taste for spatial order extended to the public space of sidewalk and street. Store awnings presented a nearly continuous canopy along the north side of the street. Their braces regularly abutted the sidewalk edge as well as the facade height. Plank sidewalks of uniform width provided pedestrian walks and a buffer between street and building. All in all, the tidy prosperity of the second-generation streetscape evinced an order and uniformity lacking in Hall's first-generation view. Admittedly, Barber and Howe's generally boosterish publication simplified and sanitized the view of the city, but the regulating lines of uniform facades, awnings, and sidewalks were hardly an artist's conceit. Factually and conceptually, the view captured the aesthetic ideal of ordered neatness that guided the remodeling of the commercial and civic districts. The tree-lined courthouse square provided a landscaped counterpoint to the hard architectural edge of the commercial district, but the two districts nonetheless shared the same underlying principles of clear, linear boundaries.

While architectural critics were urging merchants to invest in architectural improvements as their civic duty, merchants were passing regulations that, ironically, restricted their own architectural liberties in their pursuit of an ordered streetscape. During this second generation, circa 1830s to 1850s, city authorities across the state began to restrict a businessman's building and retailing practices if they intruded into the public space of the commercial district.[67] They were not replacing commercial priorities with civic ones, but the rash of municipal ordinances regulating the appearances of public space showed a more nuanced process of intermingling of the two. Certainly, there were pragmatic issues at stake. As the commercial district grew larger, denser, and potentially more obstructed, local governments interceded to keep the streets and squares open and clear for the free movement of people and goods. The overflow of suspended signs and tumbling sidewalk displays impinged on the ease and efficiency of movement in the commercial district. But there was also a cultural aesthetic at work which now valued beauty in the form of visual order and even possibly the notion of the public's, as opposed to the merchant's, right to public space.[68]

The authority of the local government did not extend to the actual design of private buildings. Yet at the plane of the street facade, the very nexus of public and private space, the city council acted on behalf of the public interest and intervened in a shopkeeper's spatial practices. Nearly identical ordinances limited the extent of structural encroachments onto the sidewalks or streets by limiting the projections of "porch, stoop, cellar steps, cellar door, cellar way, or platform" to five feet on Rochester's wide main streets and six feet on Syracuse's main streets.[69] Similar ordinances also controlled how much sidewalk space a shopkeeper could appropriate for commercial displays and trade. No "dry goods dealer, grocer, auctioneer, manufacturer or merchant, or any other shopman or dealer of any kind" was permitted to "place any goods, wares, merchandise, or other articles, in front of any store, shop or other building" farther than three feet in Rochester or six in Syracuse."[70] Syracuse reduced it to four feet in 1857, indicating that sidewalk space was increasingly considered public, not private, space.[71]

The regulations conceived of the commercial street not just in plan but in elevation as well. In Syracuse, stacked sidewalk displays could rise no higher than four feet.[72] Signs were efficient means of selling and added to the visual legibility of the district's commercial identity, yet they also were regulated as intrusions into the visual uniformity of the street. Both Syracuse and Rochester shopkeepers within the business district had to limit the projection of their signs or fixtures to two feet.[73]

The regulations indicate motives beyond convenience, utility, and accessibility—specifically, a desire for uniformity for its own sake. The specificity of the details, down to the inch in some cases, allowed for very little variation in the architectural toolbox of building appurtenances. In both cities bow windows could not project more than fourteen inches.[74] Rochester awnings must "be constructed in a uniform manner," specifically seven feet high, rounded not squared, with a minimum diameter of five inches, and set in the ground flush with the outer edge of the sidewalk in line with the curb stone.[75] In Syracuse awnings had to be at least seven feet above the sidewalk, with their posts aligned with the inside curb stones.[76] This kind of uniformity worked hand in hand with the repetitive architectural styling to create a regular, even, ordered streetscape that celebrated its own comfortable aesthetic. The second generation of merchant city builders was listening to those architectural critics who called for business interests to lead the way in introducing architectural improvements to town planning. The spatial frame of the commercial district had been convincingly erected.

Architectural beauty and rational order had been imprinted on a static cityscape, but in reality the city never stood still. The perpetual motion of Rochester's commercial district bombarded Nathaniel Hawthorne's senses in 1832: "The whole street, sidewalks and centre, was crowded with pedestrians, horsemen, stage-coaches, gigs, light wagons, and heavy ox-teams, all hurrying, trotting, rattling, and rumbling, in a throng that continually passed, but never passed away. Here a country wife was selecting a churn, . . . there, a farmer was bartering his produce; and, in two or three places, a crowd of people were showering bids on a vociferous auctioneer. . . . At the ringing of a bell, judges, jurymen, lawyers, and clients, elbowed each other to the court-house. . . . In short, everybody seemed to be there, and all had something to do."[77] The merchants' refinement of the commercial district did not stop with the physical cityscape but extended into the human landscape as well. Upon deeper investigation the people, activities, and sites that amazed Hawthorne were not jumbled within one indiscriminate public space of the commercial district but, rather, were part of a culturally scripted cityscape in action. Socially and spatially, people, too, were sorted out within the public space of the commercial district.

FIVE

Gentrifying the Sorted City
Social Sorting in the Commercial District

One Rochesterian proudly explained that the commerce in his city was "like a whirlpool which draws everything to its centre."[1] But, however accessible it may have been, this "whirlpool" was neither a chaotic nor neutral space. The same bourgeois city builders who constructed the morphological and architectural armature of the sorted city also forged social codes that reiterated their belief that good urban order resulted only when everything and everyone was in their place within the public realm. Knowing one's place, however, was not a static spatial concept. It involved the use of that space.

The rise of the middle class, including its repudiation of the lower classes, found strong spatial expression in the public space of downtown Rochester and Syracuse. During the second generation of settlement merchants and professionals commissioned showy buildings and manipulated the built environment to build economically productive and urbanistically aesthetic cities. The business rows and business blocks and the regulated sidewalks before them were not only part of the aesthetic renewal of the cities but also part of the social construction of a bourgeois cityscape. Gentrification was both an architectural and social process. The white middle class of merchants and professionals guided the physical urban landscape into a genteel social landscape that reinforced their own bourgeois class identity and aspirations.

The creation of class-based, racialized, and gendered spaces within the commercial district was both a social and physical construction of space that reflected the continued prominence of the white, male, merchant class in ordering the urban environment.[2] Mills, warehouses, basins, shops, offices, and homes displayed the personal wealth of the merchant elites as well as their literal and figurative investments in the city. The aestheticized commercial district additionally showed that the merchants' influence in the built environment extended beyond the authority over their own property and into the public realm of buildings and sidewalks within the commercial district. The social gentrification of the commercial district showed how pervasively bourgeois culture infiltrated downtown, influencing even the social expectations of a proper use of public space.

The architectural refinement and social gentrification of the commercial district paralleled the merchant's personal evolution of his own well-ordered landscape. The rise of the market economy, class consciousness, and religious revivals all inspired this new businessman to seek control of his environment. As the historian Paul Johnson concluded, Rochester businessmen "became resolutely bourgeois between 1825 and 1835": "In 1825 a northern businessman dominated his wife and children, worked irregular hours, consumed enormous amounts of alcohol, and seldom voted or went to church. Ten years later, the same man went to church twice a week, treated his family with gentleness and love, drank nothing but water, worked steady hours and forced his employees to do the same, campaigned for the Whig Party, and spent his spare time convincing others that if they organized their lives in similar ways, the world would be perfect."[3] Self-discipline coupled with an active participation in the consumer revolution became a conveniently moralized bourgeois value. The archetypal bourgeois businessman brought his convictions to interventions in the built environment, but he did not do it alone. Intermarriage between merchant families and active participation in the Episcopal or Presbyterian churches forged a powerful "federation of wealthy families and their friends" which bound together the entrepreneurial community on a common class identification.[4] This is the class whose "initial leadership in land ownership, occupational status, and religious and political office combined with education, advantageous marriages, and the perquisites of power to extend their control over local social, economic, and political domains."[5] The sorted, well-ordered cityscape was the product of their efforts.

Far from stopping at the functional and architectural sorting of the physical cityscape, the ordering impulse of the city builders extended physical ordering into the realm of social sorting. Believing that a well-ordered city was a sorted city, the merchant class melded architectural and behavioral norms to create a spatial culture that influenced how space was used in the most public of all city realms—the commercial district. From outside to inside, and cellar to attic, the functional uses of commercial space were invested with social expectations about their appropriate use. Just as the interiors of buildings became coded with social expectations, so, too, did the open streets and sidewalks. White males, middle-class females, African Americans, and Native Americans all used downtown public space, but they did so differently. Guiding the social gentrification of the commercial district was the ever-present hand of the merchant city leader and his quest to create a genteel, bourgeois urban environment.

The spatial culture of the frenetic city—"everybody seemed to be there"—which had bombarded Nathaniel Hawthorne was actually far more orchestrated than it first appeared.[6] As Michel Foucault and Paul Rabinow have argued, architecture can be a "political technology" used to exert "control and power over individuals."[7] As a physical means to a social end, architecture "contributes to the maintenance of power of one group over another at a level that includes both the control of movement and the surveillance of the body in space."[8] In the case of Rochester and Syracuse that power was localized in the bourgeois merchant elites who dominated the physical production of space as well as its social construction. Women, minorities, and members of the lower classes who were on the sidelines of economic and political power were brushed aside from figurative ownership of the commercial district.

Social Sorting in Public Space

In some senses the entirety of the new city was in the public domain. Residents and visitors traipsed over the new cities as if these overnight sensations were wide open—and in a sense they were. Gazetteers, traveler accounts, and gregarious residents took pains to describe the various buildings in town, pointing out what could be seen just by perambulating the streets. Sightseers used these sights to draw conclusions about the material and social advancement of the settlement. Not only were public buildings discussed but private ones, too, in essence making the private also part of the public sphere. These practices, according to the historian Miles Ogden, "meant that the city could be understood, and presented to the individual, as a public space open to the wanderings and gaze of the walker."[9] Similarly, Michel de Certeau reminds us that architecture and urban space were not absolute in controlling spatial experience. The very act of living— "walking, naming, narrating, and remembering the city"—could be a subversive "spatial practice [that] eludes urban planning."[10] But the new city's proud accessibility was actually guarded by members of the bourgeoisie, who construed the notion of public space through their physical production and social construction of urban culture. Being in urban space and gazing on urban sights did not equal possession or entitlement to that space. Instead, the real standard of belonging to the public sphere was the right to claim physical space and inhabit it. The constricted spatial experiences of white bourgeois women and racial minorities indicate that the popular figure of the liberated flaneur must have been a white male.[11]

In analyzing the movement and interaction of three social groups in the commercial district, it becomes clear that white males of various ages and social ranks had the greatest right to public space as befit their expected, if not executed, role in building the economy and culture of the city. At the other end of the spectrum of spatial freedoms lay the Native Americans, who were negligible producers or consumers within the mercantile economy and were, moreover, refused entry into urban society. White bourgeois females fit in the middle, experiencing spatial privileges as well as restrictions.

The consciousness of these different codes of comportment was captured by Edwin Scrantom, a member of the first white family to settle in Rochester. In 1856 he recorded his content assessment of the view outside his Buffalo Street store. The March day was pleasantly warm, the ice was thawing, and there was not an Indian in sight. As he was enjoying the urban milieu, Scrantom observed a difference between the sexes. Like himself, the men were relaxed, "standing all over in the sun." The women, however, coexisted in a different tableau—they were a "circulating plenty."[12] In the new city's idealized, ordered, hierarchical spatial culture the urban space was the same for all, but its urban practice was not.

White Males

At the apex of the urban spatial hierarchy, white men and boys had the greatest spatial options, enjoying even the most passive privilege of lollygagging about in public. Relaxing by the canal towpath that cut through the cities proved a popular past time. Erie Canal towns typically included low bridges that connected the halves of the canal-bisected community. The stone-arched Salina Street bridge in Syracuse was no exception. Its three-foot-high sidewalls were capped by a three-foot-wide coping that provided "a favorite lounging place for the lazy people of Syracuse."[13] Here boys and men waited to hitch rides on the passing boats, to steer travelers toward a particular inn, or just to watch the sights. Story has it that some loafers were known to fall asleep on the parapets and roll into the canal. The towpath also provided lounging spaces, where men relaxed in chairs pulled from their warehouse offices (Fig. 26).

A male's prerogative to loiter in public was confirmed in the outdoor porches of the prominent hotels, which were already gendered spaces associated with male business, politicking, and socializing (Fig. 33). The hotel veranda, in particular, was a democratically opportunistic space for those males who ventured out to its rocking chairs and railings. According to one resident, the Syracuse House presented two types of porch sitters: businessmen of "leisure" who gathered to "discuss the news of the day and the gossip of the town"; and lowly poseurs whose vanity "led them here that they might be seen by the people passing by and be taken as guests of the house, as they were always picking their teeth."[14] In a piece of local fiction a young mechanic seeking to establish a reputation for himself in Syracuse punctually appeared on the hotel veranda every night at six o'clock (Fig. 21).[15] As the historian Richard Bushman has explained, "Genteel spaces had immense authority because being there—at the right moment in the right dress—identified a person as genteel."[16] Intentional landmark institutions such as the Syracuse House promoted the city at the same time that they reflected the hegemony of the business interests that ran it. The porch sitters clearly understood the message of the Syracuse House and used the setting to their own advantage. Considered inappropriate for women and minorities, public loitering was elevated to an art for men.

Inside the semipublic space of shops and hotels, white males also idled, unless one considers smoking a dynamic activity. Palmer's tobacco store doubled as a Rochester political club whose paying members were entitled to smoke all they wanted of his stock provided they did so on the spot.[17] Men used a similar strategy to claim the hotels' public parlors. A hazy room filled with cigar-smoking men with propped feet was not a genteel space for women travelers, who were forced to retreat to their own rooms if no separate parlor was available.[18] A new breed of gentleman also found the scenario oppressive. Complaining to his wife about living in a hotel, Timothy Cheney of Syracuse wrote: "I have got so tired of living in this off hand way this boarding & sleeping with Tom Dick & Harry I don't believe in. I have got sick of it. . . . Our Sitting Room after Supper is on the side walk or in Some Rum Hole. I am tired of loafing about in these Dens of poison and degraded places."[19]

The interior of the Mansion House in Syracuse during the 1820s was typical of the newer hotels. Inside was the office, a sitting room, and a dining room. In the public rooms newspapers were strewn about, and the walls were covered with local and statewide advertisements regarding elections, boat and stage fares, real estate opportunities, lectures, auctions, and stores.[20] As part of the refinement of the commercial district, hotel interiors were similarly improved to provide even more plush settings for the public rooms. The fitted-up barrooms in the new city hotels were entrenched as local gathering places for the city's businessmen to drink and gossip. "In fact," one pharmacist recalled, "a large number of business men of the street made it a rule to go there every day, and it was considered eminently respectable to do so."[21] During the 1840s the Rochester Eagle Hotel upgraded its street-level rooms. The renovated first-story dining room, touted as "clean, fresh, inviting," was accessible only by passing through the saloon, whose sofas, padded easy chairs, and fancy mirrors invited public—that is, male—repose.[22] The men's refurbished drawing room on the second floor connoted luxurious domesticity. It was outfitted with velvet cushioned sofas, ottomans, and easy chairs; marble tables graced floors covered by patterned Brussels carpets, and the walls sparkled with immense mirrors in gilt frames.

Each hotel developed its own reputation, in terms of the quality of its overnight lodgings and the status of its daytime customers. Sitting virtually next door to each other, the Rochester House and Clinton House served two distinct classes of white men. The Rochester House provided the meeting space for the town's elite (Fig. 13). Augustus Strong recalled his introduction into select male culture in the great parlor of this hotel. There the Orion Club, a self-proclaimed "galaxy of stars," gathered as a debating club. The society prided itself on being spartan and distinct from working class, drinking culture.[23] It was also the home of the state's canal weigh master and toll office, an institution whose import was reflected in its fine surroundings. If they were looking for measurable class distinctions, they needed look no farther than the Clinton House, the domain of the boat captains and canal men who were drawn to its Kremlin Saloon restaurant, reasonable rates for lodgings, and convenient location opposite the Erie Canal boat office.[24]

Although it challenged the veneer of gentility, white male spatial privilege was so entrenched that even its lower-class version was nearly impossible to eradicate. Canal-side groceries and groggeries lacked the polite pretensions of a fine architectural setting or a moneyed clientele. Groceries might sell fresh food and canned provisions, but many were simply purveyors of liquor, earning the disgust of the moralizing middle-class merchant class, whose members had adopted sobriety as a businessman's moral virtue. Both Rochester and Syracuse had a swath of rookeries, known colloquially as "Chicken Row" and "Robber's Row," whose ramshackle buildings catered to the lower-class resident and canaller. A self-appointed upright citizen explained the debauched origins of the name Robber's Row:

"In former years the tenants in that Row seemed to strive with one another to see which would sell the most whiskey or 'rot gut' and get the most drunk, and rob their customers of all the money they had by them. Not long ago a dozen men were seen laying upstairs over one of these groceries, dead drunk. Hence the name of 'Robber's Row.'"[25]

Drinking was a thorn in the side of the teetotaling merchants, who tried to wrestle the physical and social cityscape into polite submission. Temperance hotels offered them a sociable retreat. But the public consumption of alcohol was a particular problem along the canal for the simple reason that this was one of the most public spaces in the city, both for residents and for travelers. When, on election day, the Syracuse schoolgirl Augusta Rann and her girlfriend tried to go to the library beside the canal, they turned back, cowed by the presence of "quite too many men around the hall."[26] Edwin Scrantom's disgust over election day drunkenness in Rochester in 1851 indicated the scene that had literally repulsed Rann: "Election all day in the City, for City Offices. Snowed all forenoon and rained all afternoon, what scores of inebriety, swearing, polluted, foul-mouthed creatures, calling themselves men, have thronged the polls and boards today, vomiting their votes and blackguardism, both polluted with liquor and bribes. Such are some of the shrines at which freedom is worshipped. Horrid desecration."[27] By 1835 temperance reformers estimated that more than fifteen hundred liquor-selling establishments lined the route of the Erie Canal, a figure that results in an average of one tavern or grocery for every quarter-mile, and an 1843 canal excursionist noted that "at almost every lock and water place through the whole rout [sic] there are from 3 to 6 groggeries, and all these for the benefit of the travelling public. . . . 'Rum, Gin, Brandy, Wine, Beer, Cider, Bread, Milk, and Groceries,' meet the eye every few miles."[28] The inebriated visions that visitors came away with were far from the boosterish impressions city builders tried to implant with their institutions such as the Syracuse House hotel, only a stone's throw from Robber's Row.

The more alcoholic groceries doubled as social centers for the whites and, to a lesser extent, the Native Americans, who could not present themselves on the verandas or parlors of the great city hotels. As one English traveler in the Midwest noted: "A grocery is . . . in fact, a dram shop; and very often is entirely devoted to the selling of spirits. . . . I stepped into one of the stores, which was full of men lying about on the counters, or sitting in chairs, balanced on their hind legs, the legs of the sitter being thrown upon the counter."[29] The activity often spilled out to the towpath or sidewalks.

One disapproving observer noted that the scene was antithetical to the well-ordered landscape: "The everlasting 'grocery' forms a conspicuous object in these villages, and bands of squalid, blear-eyed Irish rowdies, with here and there a solitary Indian luxuriating in all the grandeur and pomposity of Indian drunkenness cluster around the doors."[30] All in all, the descriptions painted a scene quite different from the urban one espoused by the emerging middle class. The socially fluid space of the grocery and towpath accommodated a variety of users, but it was an unseemly place for bourgeois men and a completely unacceptable space for bourgeois white females.[31]

Temperance efforts championed by a moralistic merchant class targeted the sites of alcoholic consumption. Bourgeois males and females attempted to rectify the situation by improving the conditions of the lower classes through judicious aid and moral example. Just as the Syracuse Company had rebuilt the Syracuse House as a beacon of enterprising urbanity to strangers, residents' efforts to clean up the city socially were also embedded in the rhetoric of city building. When a Rochester newspaper urged, "Let these waters be pure; let the canal be a proud monument to the passing stranger, as well as of our public virtues as of our commercial enterprize [sic]," it was referring to the religious fiber of the canal community, not the clarity of the waters.[32] The Erie Canal, narrated Herman Melville, "flows one continual stream of Venetianly corrupt and often lawless life."[33] Social reform promoted the city's reputation as a moral society.

By definition the canaller was not rooted to the city but, rather, bobbed in and out, "a terror to the smiling innocence of the villages through which he floats; his swart visage and bold swagger are not unshunned in cities."[34] It is ironic that the canal landscape was so shabbily encoded in the sociospatial hierarchies of the city, given that, without the canal, Syracuse would likely have remained a hamlet and Rochester's growth would have slowed. The canal gave the economy the push it needed to move its residents up into the middling merchant classes. The transient canallers and the people who served them found themselves in an odd position; they were both morally suspect yet remuneratively valuable.

Nonetheless, the moralized merchants in the cities did try to shun the canaller. In 1825 the Sabbatarians' attempts to close the canal entirely on Sundays failed, ironically, because of the implications for morality. The state legislature accepted the idea that canallers could be a disreputable lot yet concluded that, if they prohibited travel on Sundays, "vast numbers would throng the canal above and below, and many persons from on board would resort to the taverns, grog shops and houses of ill-fame, that would soon abound in the vicinity of the locks, and most of the vices which degrade and debase mankind would no doubt be encreased [sic] to a much greater extent than if the boats were permitted to pass."[35] Transient sin, they decided, was better than docked sin. By 1841 Syracuse authorities had outlawed the sale of alcohol "in any street, square, basin, canal, or other public highway" within the city under the penalty of a twenty-dollar fine. Yet the reality of the canal landscape forced a compromise. The penalty acted more as a fee for license; moreover, alcohol could be sold "from the bars of packet boats on the canals, and to be drank therein."[36]

Architecture, however, could obliterate what social pressure could not. Across the state Christian reform groups commissioned churches for the canal zone. The Rochester Bethel Church was actually built over the site of Chicken Row, replacing the rookery near the courthouse with a Greek Revival mission church that further extended the civic district along Buffalo Street. Always attentive to the prerogatives of commerce, the congregation subsequently relocated in the 1850s to Sophia Street and sold their commercially valuable Buffalo Street lot.[37] Robber's Row in Syracuse also finally fell under the pressures of real estate development. The Rochester Arcade and Athenaeum buildings (the subject of the next chapter) were a similar social reclamation project that evicted undesirables from the commercial district without actually purging them from the larger city.

In their attempt to refine the social landscape of the canal, local leaders found themselves less powerful than they had hoped. The state actually controlled the canal and towpath and thus the property within the city's municipal boundaries. As such, the frustrated city authorities and landholders were forced into the role of petitioners, requesting outside approval in the handling of local affairs. In 1830 Rochester's leading businessmen, including Jonathan Child of Child's Basin, joined with the elected municipal authorities to petition the state canal board to revoke the license it had granted for a shopkeeper to operate along the towpath next to the basin. Instead of the anticipated sturdy warehouse, he had erected a flimsy grocery. "The valuable property in the neighborhood is constantly exposed to fire [and] the citizens . . . are annoyed by mobs and collections of disorderly transient persons in and about the shops."[38] Although forced to go through legal channels, the merchants ultimately prevailed. Upon the expiration of the offending shopkeeper's lease, the permit was rescinded.

Bourgeois Females

Urban life presented a different set of socio-spatial challenges for white middle-class females, who confronted cultural messages that the ideal urban landscape was a masculine one. While the emerging rhetoric of middle-class domesticity placed women securely in the home, surrounded by the buffer of familial privacy, the reality was that women frequently ventured into town and, indeed, had great freedom in where they went. Their experience of place, however, was distinct from that of their white male counterparts, who dallied about in public spaces.

The 1819 edition of *The Whole Duty of Woman*, one of the first books published locally in Rochester, showed the disjuncture between the ideal and actual use and users of public space. The whole duty of woman, according to the author, was to be virtuous within the confines of her domicile. And yet women obviously were not, thus etiquette guides instructed women how to behave in public. "Be not frequent in the walks, nor in the thronged parts of the city," *The Whole Duty* warned; the exemplary woman "frequenteth not the public haunts of men; she inquirith not after the knowledge improper for her condition."[39] As the cultural historian John Kasson has pointed out, during the first half of the nineteenth century "segmentation of public and private life was rapidly increasing, the public arena was fraught with special concern as a problematic realm," particularly for women.[40] The bustle of city life inevitably included the random obtrusions of a male's words, gesture, touch, or gaze. An 1827 letter between two prominent Rochester businessmen painted the picture of Rochester's commercial district, a milieu unavoidably thronged by men: "Business goes on, and briskly... hammers clink, carts rattle, streetmen brawl, boys halloo, and cryers cry 'hear ye' &c. Lawyers and doctors are thick as ever. Idlers and dandies strut as usual. The theaters, museums, and pictures and other curiosities about as abundantly as formerly, and men 'in the full fruition of unrestrained liberty' pass to and fro, gathering substance and leaving 'pomp and circumstance' behind them."[41] The whole duty of women may have been to efface themselves from the masculine, urban environment, but it was an impossible charge for the urban resident.

Charitable work provided one socially acceptable excuse to be out in public. The Rochester Female Charitable Society was organized in 1822 in response to the social problems of the sick, poor, and depraved of all ages and both sexes who wandered the streets knocking on doors and importuning passersby for aid. In order to purge the "vicious" elements from the streets, the Charitable Society drew up a plan of visiting districts and assigned members to become acquainted with the neighbors in their districts to assess accurately their levels of need and to distribute aid on the society's behalf.[42] Their system of personal inquiry and home inspection remained in place through the mid-nineteenth century and was copied by other cities, including Syracuse.[43]

The domestic missionaries of the Female Charitable Society at first congratulated themselves on the efficacy of district visiting, convinced they had ferreted out the deserving from the undeserving poor. But in 1836 Mrs. Kempshall, whose husband owned a large mill on Child's Basin, resigned from the charitable society after a difficult year. There were, she explained, "whole Districts appointed to females as visitors where no decent female should go" and where they dealt with "vile and degraded inhabitants." Her district included the canal towpath, where she went almost daily to check on eight or ten families crowded together into two houses. Their depravity led her not only to abandon hope of getting loaned articles returned but made her fear that they would retaliate for any perceived indignities by burning the Kempshall's house.[44] Spurred by the resignation of fearful members, the society pushed for the creation of a workhouse where the poor would be sent to earn their keep. The subsequent construction of an orphanage and workhouse confirmed their sense of accomplishment. And, because they were confident that the deserving poor had already been succored, any street beggar was by default undeserving and could be ignored in good conscience.[45] In Syracuse the workhouse was seen as such an efficacious solution that the city passed an ordinance in 1849 which outlawed begging without written permission.[46] The bourgeois cityscape had theoretically been socially cleansed. Ironically, this also meant fewer reasons for a woman to have business about town.

During the same period local authorities tried, with spotty success, to legislate decorum within the urban spaces under the corporation's purview.[47] To make the streets respectable, city ordinances sought to control indecent language and indecent or disorderly conduct in public. The new cities passed almost verbatim the same social policy ordinances. The transgressions included "any noise, disturbance or improper diversion" in the streets and squares of the city "to the annoyance or disturbance of citizens or travelers."[48] The extent of legislation, however, extended beyond the predictable rules against breaking the peace. City councils and village trustees further tried to legislate decency. Targeting the public nature of public space, they proscribed vulgar, profane, and obscene language and conduct in any street, street corner, bridge, or public place, including the market halls that shouldered the commercial district. Even the common male privilege of swimming naked in the canal was prohibited during daylight hours. The bourgeois leadership both instigated and enjoyed the improved landscape. Their wives and children appear to have been the targeted beneficiaries of these morally uplifted streets, particularly those that were most intensively and commonly used, that is, those in the commercial district. The mention of travelers in the ordinances maintained public pressure for a genteel decorum that would reflect favorably on the brash new cities and possibly entice the right kind of gentrifying settler.

Middle-class white females developed a separate spatial culture within the shared physical space of the city. They shaped their urban environment through their own comportment. The way that they dressed, behaved, and moved while in public created a real but invisible female space—a mantle of private respectability—which made the city available to them. One nineteenth-century guide drew an analogy between manners and a fortress: "[Etiquette] is like a wall built up around us to protect us from disagreeable, underbred people who refuse to take the trouble to be civil."[49]

Although separated by thirty years, both *The Whole Duty of Woman* and *True Politeness for Ladies,* also circulating in upstate New York, agreed that women should build genteel walls around themselves so as to be as unobtrusive as possible in public. Proper comportment, *True Politeness* explained, required subdued conduct, dress, ambulation, and gestures. It urged ladies to dress plainly when going in public and reminded them that a dress worn for walking required a different style, material, and ornament than one worn for dinner. In the likely case that a woman encountered men on the streets, *Whole Duty* instructed her not to turn her head "to gaze after the steps of men" or to be so bold as to inquire of them where they were going.[50] *True Politeness* established a hierarchy of greetings in order to avoid social gaffes. It reminded ladies that "the superior in rank and station should first salute the inferior" and that, by extension, "if you meet a gentleman in the street with whom you are acquainted, recollect that it is your province to recognize him before he assumes to salute you."[51] Should an inferior presume to salute a woman before first being acknowledged, he or she was to be gently ignored under the pretense that the lady supposed the greeting was intended for someone else. Any man or woman who would gauchely presuppose the right to be acknowledged on the street based merely on a previous introduction at a ball or tea was to be set straight with either a cold bow or, better yet, a complete lack of acknowledgment. As the guide explained, an introduction at a ball or at a friend's house "does not compel you to recognize the person in the street," nor, to be fair, does it "entitle you to future recognition by such person."[52] *True Politeness* was emphatic that under no circumstance was a lady to "boisterously salute" or, worse, shout out the name of a female acquaintance in public. To do so would compromise the privacy and decency of both ladies.[53] Such advice was seconded in the training that Miss Araminta Doolittle gave her students at the Rochester Female Academy. Her student Alice Hopkins later recalled, "What she wanted was to make us all over into high-bred, courteous, cultivated, truthful women of society, well-dressed and, above all, without eccentricities, trained never to do anything especially to attract attention."[54] The ideal lady was an invisible one.

Purposeful walking was critical to navigating the public landscape. As suggested by the terms *street walker* or *public woman*, such rules of conduct were needed to permit a genteel woman to be out in public.[55] Moralizing literature painted a picture of immoral women defined by the streets. The sentimental paean "Hymn for Female Penitents" published by a Troy authoress portrayed the fallen woman by her rambling habits: "Much hath she sinned—for many years / hath walked, by night, the city's street."[56] In the novella *Life in Rochester* Chumasero chose the suggestive name of "Eliza Streeter" for a naive girl ruined after becoming the lover of a man whom she had met on the street.[57] Phebe Davies, a seamstress of Syracuse, was sent to the Utica Insane Asylum by the orders of the county sheriff on the charge that she was a dangerous person to the community. Because her story is told by Davies herself, her transgressions are unclear, although it is indicative of her unconventional habits that when the authorities seized her she was found on the street "walking out for the benefit of my health."[58] By 1857 Syracuse authorities passed an ordinance directing that any woman "found loitering or strolling about the streets of the city, by day or night, without any lawful business" be fined ten to fifty dollars, the same fine charged to a convicted prostitute.[59] Etiquette guides stressed the importance of continuous movement while on the street. *True Politeness* specifically recommended bowing, not curtseying, when acknowledging acquaintances, because that protective form of greeting would not interrupt the flow and grace of the stroller's forward motion.[60] Appearances were everything; females had great access to public space but only if they kept moving.

The Syracuse teenager Augusta Rann adopted just such a continual "maneuvering" as her spatial strategy.[61] She was an intensely peripatetic girl, whose walks took her not only through the Syracuse commercial district but also miles up to the Salina saltworks and over to the Geddes Idiot Asylum. While out on the street, even if simply promenading, Augusta took care to appear as though she had a destination in mind. Far from loitering, she was not available for conversation or even an exchange of glances. Her diary is peppered with accounts of ignoring people, especially male classmates, while out on the street.[62] "After school went to the bookstores with Miss Lathrop . . . just as we were passing Johnson's grocery who should I see but J.M. I did not let him know I saw him, for fear he would think I came down the street just to see him." The characters might change, but the avoidance techniques remained the same. "After school went uptown to the P.O. and to the library with Gertrude King. Albert B waited on the stairs until after I came down, but I did not look at him, he went along before us, when we went uptown, and he looked back pretty often." Young Rann was, of course, looking around while trying to appear not to. "Went up throughout the city with Orissa Roach saw no one worth seeing," she complained more than once; "Promenaded a long time, but saw no one else."[63]

It would be wrong to ascribe young Augusta's eyes-forward, feet-moving approach to enjoying the city strictly as an adolescent behavior. Her spatial strategies were not only sanctioned in the bourgeois manner books but were echoed in older women's comportment in the city as well. Alcesta Huntington of Rochester similarly hustled about town as a teenager and an adult.[64] The stakes could be high. When Sarah Littles was accused of murdering her husband near the Rochester falls, she presented an alibi showing she was a reasonable and decent woman who, incidentally, was nowhere near the scene of the crime. The evening in question Sarah had walked widely through the city, passing in and out of several urban districts, ranging from her mother's house to the south, to running errands in the commercial district in the center, heading toward a friend's room in the milling district to the north, and back home again through the commercial district.[65] Throughout her stroll—with one exception—the irreproachable Littles always had a destination in mind and thus a purpose to her walk. The one time she seemed aimless was indoors, and even then it was unacceptable—the dressmaker Mary Farrell threw her out of the shop for loitering.[66] The purposeful rambles of Rann, Huntington, and Littles all point to the relative freedom that white middle-class females enjoyed in the city. Their behavior during their rambles, however, points to a common strategy of using movement to maintain the mantle of privacy when on a public street. While out in public, a lady must appear to have a goal and must keep moving.

Social mores discouraged women from sitting out in public, where it would be difficult to avoid contact or observation. If a lady wanted to stop, she had to find a proper retreat from the public. A foot-sore Augusta Rann sat with her aunt in the women's parlor of the train station as they awaited her uncle. Alcesta Huntington buried her head in the decorous *London Industrial News* in the second-floor reading room of the Rochester Athenaeum.[67] The space, the seriousness of the reading matter, and her posture buffered her from most public scrutiny. Reading traditionally provided a genteel illusion of busyness. A Syracuse resident recalled that during the 1820s the Syracuse Book Store "was the headquarters for the better class of village loungers, the intellectual folks in their idleness."[68] In terms of genteel respectability, sitting in a hotel's public parlor was nearly as bad as lolling about its front veranda. During the 1830s one Troy mother worried greatly about her young daughter's comportment while traveling through the state: "She is making *herself* very *conspicuous* . . . she is very young and wants a great deal of council. I hope you will never let her be in the public parlor without you or Harriet are with her, I am afraid that she will get to be a forward girl, and that I could not endure."[69]

During the refinement of the new cities during the 1830s and 1840s, hotel owners began to carve out gender-segregated spaces that delineated the public-private, male-female distinctions.[70] For example, after the Rochester Eagle Tavern was rechristened the Eagle Hotel, it was upgraded to provide a ladies' parlor on the second floor, safely above the men's first floor bar, dining room, and a courtyard smoking platform, and far from public scrutiny.[71] These designated women's parlors were semipublic spaces intended as bastions of sequestered and passive female entertainment. One young traveler noted that all hotels had rocking chairs placed out on the large wooden verandas, but she herself sat indoors in the parlor overlooking the bustle on Rochester's streets. The Syracuse House hotel impressed the traveling Miss Leslie in 1845, who noted its high style and fine accommodation to privacy: "While in Syracuse in 1845 Miss Leslie took apartments at the Syracuse House, a very spacious and very fine hotel at a corner of the great square. A large portion of this house was so arranged as to give each guest a commodious parlour with a small chamber opening into it—a most excellent plan. Those parlours (of which I had one) were all very handsomely furnished in city-like style; and the bedrooms were light, airy, and of comfortable size. The drawing-room opened on the balcony, from which was a fine view of the square, with the canals and bridges."[72] Indeed, a resident's engraving of the Syracuse House showed women ensconced in the upper balcony while the men gathered at street level (Fig. 20). Far from posing on the hotel's veranda, genteel ladies modestly withdrew.[73] Although Michel de Certeau and Walter Benjamin suggest that the simple acts of "walking, naming, narrating, and remembering the city" are liberating spatial practices that permit the individual to claim space, the navigational strategies of white bourgeois women show constricted access even as they walked and looked.[74] Not only were ladies expected to control their level of engagement with the public and public space, but they were also expected to regulate the public's physical and visual access to them.

Native Americans

Not surprisingly, Native Americans were at the bottom of the social and spatial hierarchies in the public space of these new cities. For all the white rhetoric of civilizing the savage, no one meant for it to happen on the white man's turf, including the public streets. Native Americans were typically presented as having no place in the white villages and cities. And yet, no matter how it dismayed many whites, Native Americans were part of the early-nineteenth-century cityscape.[75]

Separate spatial cultures divided Native Americans and whites and clouded their ability to understand the way the other used city space. As one white Rochesterian saw it, just coming to town was their first failing. "These indians were completely demoralized," Edwin Scrantom wrote; "they refused to range the forests with their wandering brethren."[76] What was worse to him was their urban comportment in the city. They dressed poorly, entered private buildings without knocking, and, finding no welcome in most public buildings, took their business outside, selling goods and even eating on the street.[77] An illustrative instance of spatial miscommunication was a French traveler's interpretive error. Finding a papoose slung on a tree in Rochester in 1818, he supposed it to be abandoned and rescued the bundled child from its perch and headed back to his lodgings at the Ensworth House. A commotion ensued as the father and mother materialized out of the undergrowth where they had been eating and demanded the return of the child.[78]

Urban spatial practices were, of course, only part of a larger culture in which the two groups collided.[79] The question asked by the white James Hall could have as easily been posed by a Native American: "How shall we deal with a people between whom and ourselves, there is no community of language, thought or custom—no reciprocity of obligations—no common standard, by which to estimate our relative interests, claims, and duties?"[80] As the historian Bernard Sheehan has pointed out, the abstraction of savagism reduced the irreconcilable differences of the Native Americans into a comforting formula. Savagism, he adds, raised "a barrier against understanding [and] set men at odds with reality."[81] The spatial culture of the Native Americans challenged the sense of order and propriety of the whites.

A Seneca ceremonial site just west of the Monroe County courthouse became memorialized by white Rochesterians as a benchmark of social progress over savagism. In 1813, on the fifth day of a nine-day Iroquois ritual, the Seneca strangled, burned, and ate one or two white-furred dogs as an act of purification.[82] Some acculturated Iroquois linked the white dog sacrifice to Christian theology. One Oneida defended that eating the dog flesh "was a transaction equally sacred and solemn, with that which the Christians call the Lord's feast. The only difference is in the elements, the Christians use bread and wine, we use flesh and blood."[83] But white New Yorkers were having none of it. The sacrificial act confirmed the white's view of the Seneca as savage and was the single event of the nine-day festival on which the whites focused. Although the Iroquois ritual persisted elsewhere, after 1813 they never held it in Rochester again. The merchants' construction of new commercial buildings and a canallers' mission church on Buffalo Street commandeered this ritual site and neatly advanced their goals for a socially and physically ordered cityscape. It was with great satisfaction that local chronicler Henry O'Reilly recalled that "the wild spots where these pagan rites were performed only twenty-six years ago has been transformed for the purposes of civilized man, and is now surrounded or covered by some of the fairest mansions and the noblest temples of Western New-York."[84] Building over the site was one way to make the ceremony, and by extension the Indians, disappear from the urban landscape.

The idealization of separate spaces was drawn in Barber and Howe's 1841 illustrated gazetteer. In the foreground of the Rochester Buffalo Street view stood the shops that had leapfrogged past the courthouse, but the caption made no mention that this had even been the site of the ceremony, nor did the section acknowledge a historical or contemporary Indian presence (Fig. 16). In the 1851 edition the authors added a new, sensationalist illustration of the pagan "Indian Worship" that had occurred on the same spot on Buffalo Street.[85] This added section symbolically sequestered the Iroquois to the back of the book. Barber and Howe similarly erased the Onondaga from Syracuse. There were no signs of a Native American legacy, let alone presence, at the hub of Clinton Square (Fig. 17). Instead, the Onondaga were isolated to their own illustrated section on the contemporary Onondaga Reservation.[86]

Although they had been dispossessed from the land, Native Americans continued to be a presence in the cities. Whites interpreted Indian behavior in town, however, without considering the effects from that dispossession. Many descriptions of the local Indians simply suggested that they were an improvident lot. In 1826 a Frenchman scorned "in the vicinity of Rochester dwell some miserable Indians who could raise an abundance of food but prefer to neglect their fields and beg at the door of every household."[87] Such accounts described the outcome but not the cause of white infiltration into Native American lands and practices and, in doing so, damned the Indian for their inappropriate response to a changing landscape. In his 1819 address Governor DeWitt Clinton noted the difficulty of the Indian situation in New York State. Based on his observations that the closer Indians came to whites, the more they "receded from virtue" or even died, Clinton concluded that "their departure is essential to their preservation."[88] Clinton's view that Native Americans were neither appropriate figures in the urban landscape nor improved by contact with white urbanization was repeatedly echoed in the recollections of the Native Americans the settlers had seen in town.[89] Whites increasingly saw Native Americans as an exotic and debased, yet safely dying, race.

Drunkenness was a recurring trope.[90] Edwin Scrantom invoked the memories of a "gang of vagabond natives" who were a frequent sight in Rochester, typically drunk and laid out on the pavement with silent, mortified wives sitting sentry beside them.[91] These sights were not tucked away, far from the public eye and daily experience of the white residents, but were, rather, part of Rochester's most public streets in the commercial district. Scrantom proceeded: "I have seen such a scene on Exchange street. . . . I have witnessed such scenes on Buffalo street. . . . I have seen them repeated oftener in a low place . . . on State street [and] Back of the Arcade was the great place in early days for Indian 'drunks.'"[92] The Indian transgressions were doubled in such a public scenario. Rochester had its share of white drunks, but they usually found comfort in the privacy of grog shops and were typically only arrested during the night, either when they were caught passed out on the streets after closing hours or when they were ejected from taverns for unruliness.[93] The lack of accommodating interiors contributed to the Native Americans' ungenteel uses of exterior spaces.

Furthermore, the Indian wives sitting sentry were forced into humiliating public scrutiny. Scrantom observed: "I have seen the squaws manifesting the deepest mortification for their condition, and their feelings finding vent in long drawn sighs, as they sat near their prostrate relatives with their heads bowed and covered from sight in their blankets. To go near a squad of prostrate Indians, stupefied with drunkenness, and stand and look at them, was to inflict great pain and uneasiness upon these patient squaw-watchers."[94] The compromised women's attempts to fend off the gawkers exacerbated the cultural dissonances. Unlike the white females who were instructed to freeze leering offenders with icy indifference, the Native American women showed their displeasure "by many signs of the head and hands; the most potent of which was the sudden turning of their hood faces towards their intruders, and then turning their backs upon them, as they took a few steps forward."[95]

Indian drunkenness threatened the bourgeois social order in other ways as well, by undermining the prerogatives of the white merchant's role within drinking culture. The semblance of social equality in which workmen drank with their employers actually reiterated the dominance of the employer, since it was done at his bequest, at his workplace, as his treat. Indian drinking fell outside this socially sanctioned form of class-affirming leisure. After temperance took hold in the gentrified community, drinking culture changed in ways that further degraded the Native American's social reception. Drinking was becoming politicized as part of working-class culture in opposition to the middle class. Native American drinking thus underscored their cultural distance from the abstemious bourgeois merchant class.[96] In 1849 a Syracuse ordinance was passed prohibiting the sale of intoxicating drinks to any "Indian or squaw, apprentice, servant or child."[97] This ordinance notably omitted the white merchant elite, thus signaling the hierarchies of alcoholic space and privilege in the gentrifying city.

By midcentury the increasingly rare accounts of Native Americans in the cityscape more often described Indians who capitulated to the bourgeois hierarchies of urban space. Far from claiming space for themselves, they made but a fleeting figure upon it. As one traveler described it, "the streets of Rochester were animated with buyers and sellers . . . and, amid the crowd of the European race, Indians might be seen in their white blankets, and with their uncovered, long, black, shaggy hair, passing in and out of the shops."[98] Hardly fixtures, these people passed in and out of the shops and kept on the move in what was probably a hostile environment. Much like the bourgeois women, Native Americans had adopted a similar spatial strategy: they kept moving.

And yet, even in their movement, the Native Americans were still seen as derelict in their comportment. Far from having the purposeful gait of a proper girl running errands, they were typically described as straggling. Mrs. Elisha Sibley recalled the Indians she saw during the earliest days of Rochester as "straggling bands and hunters [who] were constantly passing through the woods about us."[99] A generation later the same perspective was echoed in Thomas Wharton's 1830 description of Syracuse: "along the Towpath were straggling groups of Indians of the Oneida tribes."[100] Another traveler commented on the Oneida she saw shortly before reaching Syracuse: "the last remnant of the once powerful tribe of Oneidas is yet lingering in this neighbourhood."[101] The straggling that whites observed may well have been the result of the Native American's spatial disenfranchisement; there were few places to pause in the city. Alvin Fisher's 1845 painting *Remnant of the Tribe* illustrated the observations of one Erie Canal traveler near Syracuse: "We have passed several squads of . . . Indians carrying baskets, brooms, hunting apparatus, &c. I could not but think of their once numerous hordes, now no more, save a few scattered remnants of their wandering tribes, having scarcely a spot which they can call their own."[102] Whereas white males pulled out benches in front of the groceries for their own comfort, the Native Americans avoided making the same spatial claims on the public path. As befit an unwelcome visitor, they kept in motion. The spoken and tacit standards of proper urban comportment created standards that marginalized Native Americans and sorted them from the white landscape. Loitering was a privilege that extended only to white males; all others needed objectives and destinations.

There were few exceptions to their spatial disenfranchisement. Despite the mercantilist aspirations, currency was in short circulation, and merchants were forced to trade in goods, not cash. Periodically, Native Americans provided unusual relief. Annuity payments, issued by the federal government for Indian land concessions, put hard cash into Native American hands, making them suddenly welcome customers in a cash-strapped community. One early Syracusan remembered, "on pay day it was almost impossible to get inside the store for the crowd of Indians and squaws who brought their government money for him to take care of. At these times the numerous papooses, strapped on frames, leaned up against the store front, much as bulletin boards do now."[103] The welcome wore out when the money was gone.

Another exception involved the agreement between Native Americans and whites that there was money to be made in the commercial district by torquing race relations to mutual advantage. In Syracuse the Onondaga women and children carved out a special niche in the town fabric at Phinney's Museum on the edge of Clinton Square. Phinney and the Onondaga realized the commercial potential of his Indian visitors. He permitted the Onondaga liberal access to his museum because the sight of the women and children "gayly dressed, with scarlet blankets, feathers, beads, and trinkets" enjoying the music and displays created its own attraction. "The sight of them in the windows and about the buildings draws strangers to enter, for the sake of seeing them more at leisure than they could do passing in the streets."[104] In turn, the Onondaga pandered to their audience, dressing brightly and picturesquely to gain tips from the visitors.[105]

The exclusion of Native Americans from the urban landscape was both a technique and proof of the merchants' dominance in framing urbanization in terms of their own economic and cultural colonization. They repeated that the best proof of the Indians' loss of entitlement to the land itself was the fact that they had enjoyed first rights to the land but had failed to do anything with it. In contrast, white settlers had constructed bona fide cities with active commercial and manufacturing economies. This "myth of the second creation" meshed neatly with the culture of improvement which dominated the mercantile settlement of the hinterland.[106] Such a theory was argued by J. C. Myers, who traveled through Rochester and Syracuse in the late 1840s.

> When we reflect on these highly cultivated regions, bespangled with the most flourishing cities, towns, and villages, whose foundations were laid by people still living, and which region already numbers a population greater than the whole of the aboriginal hunting tribes, who possessed the forest for hundreds of miles around, we soon cease to repine at the extraordinary revolution in the history of those tribes, however much we may commiserate the unhappy fate of the disinherited race.—Because here now the noble enterprise of the white man has so changed the aspect of this region, that upon every hand attractive beauty meets the eye; and here now far and wide the aboriginal forest has lost its charms of savage wilderness, by the beauties of cities, towns and villages, and the intrusion of railroads and canals.[107]

By midcentury the suppression of Native Americans in the city permitted white residents to recall the Indian presence as a somewhat colorful element of everyday life. Indians dramatized the efforts of the pioneers, and their eradication was a measure of urban civilization. Fifty years after his family moved to that first log cabin, Scrantom wrote, "the transition certainly is wonderful, from a 'Howling Wilderness' with one log hut, surrounded by Indians,—who are always as uncertain as wild beasts, and more to be dreaded,—to a city of fifty thousand inhabitants."[108]

Spatial privilege clearly depended on one's race, gender, and class. White males created their own space in the public realm, both casually, by dragging out benches to the towpath for impromptu drinks, and officially, by constructing emporiums for business as well as verandas and saloons for male sociability. The walking of white women and Native Americans throughout the cities, and especially the commercial district, did not result in any equivalent authorship of space that white males enjoyed. Their claims upon space were fleeting and therefore weak. The physical, public spaces allocated to bourgeois white women were shaped not by but, rather, for them by white male society, and those spaces were either ephemeral or secluded. Native Americans had even fewer spatial options. It would be romantic to hold the notion that the "gaze" of these minority flaneurs was enough to liberate them from the governance of the social construction of urban space. Rather, their very exclusion from full participation in public space itself reiterated the power of those who excluded them. Certainly, each social group had its distinctive spatial practices that complicated the meaning of public space, but the heavy hand of bourgeois coding colored spatial practices and social reception.

Social Sorting within Commercial Buildings

The ordering impulse that had guided the architectural refinements of the commercial buildings extended to social ordering inside as well. Although it was not regulated by municipal ordinance, the ever-present hand of the merchant guided the process of building consensual spatial norms that distributed below-ground, street-level, and upper-story tenants and clients. Men and women, consumers and producers, whites and people of color, were virtually slotted into particular spaces and roles within the new commercial blocks. By going inside the buildings and settings of the commercial district, we see the extent to which architecture ordered people in space and framed social expectations. Being semipublic spaces, these shops and offices were theoretically open to the public, an expectation compounded by their very presence in the commercial district. But, much as the district's sidewalks and streets were commercially conditioned public space, so, too, were the shops privatized public space.

Cellars

Cellar establishments were largely the territory of men and boys. Physical, economic, and cultural considerations all played into the gendering of these below-ground quarters. Many cellars were simply dank storage facilities encountered by stock boys such as the fictional William Brown, who was described in the 1848 novella *Life in Rochester*. Instructed by his Exchange Street employer to "put on a 'tick apron, and go down into the cellar with a basket, and clean up the rubbish on the bottom," the dispirited clerk "stooped down at the bottom of the damp, earthy cellar, and scraped together fragments of hoops, decayed bits of boxes, broken bottles, and mouldy wisps of straw."[109] Other cellars were converted to public businesses, with varying degrees of finish. Proprietors typically chose cellar shops for their cheapness. Rochester's D. H. Ray's barbershop, the Kremlin Saloon, and a private employment agency (called an intelligence office) all needed to be near the customers circulating through the commercial district, but, as the providers of services, not the sale of goods, they did not need to entice customers with displays. Nor did they court female customers where decorum would require architectural amenities within a public setting. In fact, barber Ray advertised that he would attend to ladies in their homes rather than expect them to descend into his shop. Wholesalers also operated from cellar locations, where male clerks waited on male customers in simple settings. The availability of goods and the efficiency of shipping, not fashionable or ephemeral over-the-counter experiences, mattered to jobbers.[110] An advertisement for a Buffalo Street business almost incidentally included the image of a respectably top-hatted, top-coated man coming up out of the cellar shop next door (Fig. 34).

The formal distinctions of gendered retail space were clearly rendered in a midcentury advertisement for Case & Mann's dry goods store on State Street, immediately north of the Four Corners. The owners architecturally and spatially created two distinct business environments on the double-wide lot, and their advertisement showed both retail and wholesale operations (Fig. 35). In what was clearly the cellar, with a stair descending and no windows visible, the wholesale department was well suited for business. The walls were lined with stocked shelves, and display cases encircled the cast-iron columns. The interior was not glossily furnished. Its lighting fixtures were plain, there were no ceiling decorations, the floor was strewn with crates and boxes of goods, and there was no seating. Here in the basement utility prevailed where all the customers and clerks were men. In contrast, the retail department on the first floor was an airy space. A row of fluted cast-iron columns ran down the length of the shop floor, supporting the high ceiling above. Gaslight chandeliers hung from ornamental rosettes. The sidewalls and the base of the columns were built up with substantial rows of shelves and cases holding the store's wares. Counters and stools provided comfortable seating for the ladies, who were waited upon by male clerks. Presumably, large plate-glass windows illuminated the scene from the outside.

At the other end of the cellar spectrum were the seedy cellar establishments—a type that left little evidence in the historical record. One source for the nineteenth-century perception of such places, however, comes from local authors. In the popular pulp press authors developed stock "places" just as they developed stock characters. In these fictionalized settings basements contained things dark, perilous, cheap, concealed, and generally unfit. The big showy business block, one local author warned, could be misleading. In John Chumasero's *Mysteries of the Rochester* a group of debauched young men, drinking, gambling, and planning crimes, gathered there "in one of the basement stories of a lofty store on ——— street, in a room well guarded from intrusion being the back part of a recess, designed especially for the 'exclusives,' and bearing upon its door the ominous and impolite word 'private.'"[111] Similarly, in his *Life in Rochester* Chumasero set the course for the hapless Eliza Streeter's downfall to prostitution when she innocently entered an intelligence office in a Buffalo Street basement.[112] As an attorney and Monroe County judge, Chumasero was in a unique position to evaluate the foibles of human nature. As a resident of Rochester, he was able to situate these characters into familiar urban spaces.

Street Level

The dry goods rows that Syracuse and Rochester merchants forged provide a quick introduction to the functional, architectural, and social sorting impulse that underlay the construction of the new nineteenth-century cities. Functionally, dry goods merchants clustered within the commercial district to form a subdistrict dedicated to selling fabric, sewing notions, shawls, and household goods. In Syracuse of 1844, for example, of the thirteen dealers who dealt only in dry goods, ten were located on Salina Street.[113] A generation later a Syracuse newspaper reported that the confluence of transportation options created "peculiar advantages offered to dry goods dealers that the latter are centering here, and, in consequence, there is considerable competition, though rather healthful to the people than otherwise as yet."[114] The evolution of dry goods rows reflected the merchants' understanding that the benefits of a critical mass of retail customers outweighed any sales lost to the nearby competition. It also indicated the identical valuing of pedestrian, horse, and canal traffic to the totality of retail and wholesale operations.

Architecturally, merchants competed through the construction of large and stylish stores. Responding to the proliferation of dry goods selling in the area, around 1850 Colonel Vorhees extensively refurbished his Empire House hotel on the corner of Clinton Square along with the attached Empire Buildings on North Salina Street.[115] The cupola-capped four-and-a-half story Empire Hotel, with its enticing rows of uniform shop fronts, was converted to advertising copy by one of its dry goods tenants (Fig. 24).

The "great Empire" with lofty spire
Towers towards the skies,
Her wide spread wings, to the breeze she flings
Her name o'er earth it flies.
Her spacious halls and corridors
The strongest nerve will charm,
In richest taste and elegance
She's carrying off the palm.[116]

Within these ornate new buildings merchants incorporated elements of an ordered aesthetic, luxurious domesticity, and public transparency which particularly sanctioned them as public spaces fit for bourgeois women.

Merchants appropriated the emerging rhetoric of domesticity which had placed women in the private home and converted it to public commercial practice. Store "parlors" featured patterned carpets on the floor, framed pictures on the walls, upholstered furniture, and elegant gas fixtures suspended from plaster ceilings. The inflated domestic scenario included families of men, women, and their accompanying children. The Rochester Music Store was so decorously outfitted that it was used for the traditional purpose of a parlor for the untraditional reception of Tom Thumb and his bride (Figs. 35, 38). The prevalence of shop interiors, as opposed to facades, in midcentury advertisements reiterated the growing importance of the setting as much as the product in selling the idea of commerce. In many images the stocks are obvious yet subordinate to the furnishings, space, and refined characters that sanction the tableau. These domesticated spaces were touted as respectable, indeed plush, commercial parlors that became sanctioned spaces for females downtown.[117]

Instead of being private parlors, however, the commercial "parlor" required public transparency. The large plate glass windows displayed not only the goods but also the interactions inside. In 1850 a Rochester landlord advertising shops for rent in his Emporium Building drew attention to the plate glass windows, which "furnished light that is hardly to be surpassed," and to the large interior mirror, which reflected "to the front street what is going on both above and below."[118] The public's visual consumption of the goods began at the sidewalk, but commercial enticement was only part of the reason. The sheer visibility of the interior protected it from the negative associations that coded the darkened and secluded cellar businesses as unsafe, or at least improper, places for genteel females (Figs. 35, 36). Visibility sanitized social interaction and thus permitted personal contact among the mixed-gender, mixed-class clerks and shoppers. Clarity and visibility of the setting, the goods, and the people were paramount to respectability.[119]

Night lighting further sanctioned public space for female use. During an evening ramble through Syracuse in 1845, a female traveler and her companions were struck by the number of women out and about: "the chief streets presented a long line of light from the brilliancy of the store-lamps, the brightest I had ever seen. We saw numerous ladies engaged in shopping in these well-lighted stores; preferring, I suppose, for this purpose, the cool of the evening."[120] The light within and without the stores was part of the genteel setting of visibility.

The protective scrutiny of visibility was also invasive as it pushed both the shoppers and the clerks into the public sphere of commodified observation.[121] The pressure of continual performance prompted complaints from retail clerks:

> The modern spirit of competition has induced a numerous class of tradesmen to adopt a plausible but fictitious appearance of traffic—a practice which, we may readily suppose, does not diminish the hard lot of assistants. No leisure moment, consequently, must be devoted to other than the business of the shop—no intervals of rest are permitted in the absence of persons to purchase. An *appearance* of business is enforced; the hurry and bustle of a thriving trade is exhibited; in lack of other duties, articles must be packed and repacked; ribbons again and again rolled—every specious means, in short, is put into operation to impress the public with an opinion of extensive traffic.[122]

Freeman Hunt, however, concurred with the practice: "put on the *appearance* of business, and generally the *reality* will follow."[123] People and goods dressed the stage artfully prepared by architecture and furnishings.

The shop floor was a social and spatial web of male and female, white and people of color, and wealthy and working-class relations. The relationship between the male clerk and female customer was a socially sanctioned form of contact but not one without its own perils. Clerks ingeniously or disingenuously found ways to entice the customer to purchase more. Larry Jerome, a Rochester clerk at Wilder & Gorton (the predecessor to Case & Mann's dry goods store) was remembered as a wily fellow. Recalled one friend: "I remember one day when a lady whose husband had just died came into the store to buy some mourning goods [and] Larry waited on her and he very sympathetically asked her about her recent affliction. While the poor women was telling him she naturally began to cry. Larry burst out into tears, also, and the two of them wept together all the time the woman was in the store, and by the way, Larry sold that woman twice the amount of mourning goods she would have needed if all her family had died at one time."[124] Such sales tactics, coupled with the plushness of Gorton & Wilder's fancy dry goods store, contain the seeds of the turn-of-the-century department store environment, which, as the social historian Susan Porter Benson has shown, "confronted the customer with a dazzling array of merchandise in a setting designed to break down her resistance to spending money and to exploit her sense of her class position and personal attractiveness."[125] Perhaps these bourgeois perils helped redeem the idea of female shop clerks. Writers reconciled themselves to women shop clerks by conceptualizing certain businesses and tasks as more female than male. The press advocated female shop clerks, particularly in retail dry goods stores, on the grounds that women were better at the tasks of tasteful folding and conciliatory and polite conduct to customers, and superior in all matters of taste in dress. "Measuring off calicoes and tape is too light a task" for men, the article concluded.[126]

The dry goods emporiums created the misleading semblance of a broad welcome and actually implemented measures that privileged the bourgeois shopper over others. The urban historian Gunther Barth has described the emergence of the department store as creating an egalitarian space that "opened up the possibility of equal access to consumption," including the visual absorption of the displays, if not actual purchase, of the goods.[127] But architectural, financial, and social practices within stores tempered one's welcome. The first-floor retail space at Rochester's Burke, Fitzsimmon,

Hone & Company was sorted into increasing degrees of luxury the farther back one went. Anyone might stroll in for an inspection, whereupon a shopper would be greeted with bolts of prints and ginghams in the front, but she would have to be comfortable running the gauntlet of clerks before luxuriating in the high-cost silks protected in the back. Similarly suited up for a dry goods store, Rochester's Emporium Building featured a fifty-foot gallery "for a shawl or fancy goods room" accessed by two flights of impressive semicircular stairs. The mezzanine gallery literally and figuratively elevated the more precious stocks above the rabble and casual touch.[128]

Cash could be the great equalizer but only if one had it. The Syracuse Empire Block merchant at the "Red Sign" seemed to call all types of customers: "Fall has come; Winter is coming, To the 'Red Sign' all are running," and specifically beckoned, "Walk in ladies . . . Come, Farmers." He promised more than a shop filled with goods to please a variety of customers. His doggerel seemed to promise a great democratic opportunity of shopping for a diverse population, provided "your pocket now with cash is filled."[129] This Syracuse merchant ran a cash business: not credit, not trade, not barter. As the visitor Lois Freeman rhapsodized, whatever one desired could be found in Syracuse, that is, "every thing for Money."[130] The one-price, cash system had gained currency as a sales tactic in the large metropolitan dry goods stores such as A. T. Stewart's in New York City and Wanamaker's in Philadelphia as a management tactic to cope with the quantity of customers and questionable price-setting skills of the salesclerks. It was also a system that benefited customers who were part of the cash economy, typically white, urban, and middle- or upper-class.

In theory the open-front retail businesses were open to all, but a closer investigation shows that the commercial modus operandi of shopkeepers within their storefronts were not as transparent as their inviting glass windows. Architecture combined with cultural practices to create commercial spaces with varying degrees of public reception and sorting.

The absence of a one-price system in most stores meant that shopkeepers set a price based on their reading of the customer—one's shopping experience depended on the legibility of one's social status (Fig. 37). In *Life in Rochester* Chumasero bared the tricks of conniving storekeepers. In his quasi-fictional exposé, the good-hearted William Brown is apprenticed to the dastardly Swindlem Skinflint, a provisioner on State Street. There Skinflint instructs the fifteen-year-old in the art of shaving due bills, or making a profit by discounting the scrip that was a common form of currency among the mechanic and manufacturing classes of the city. Skinflint

plained, an introduction at a ball or at a friend's house "does not compel you to recognize the person in the street," nor, to be fair, does it "entitle you to future recognition by such person."⁵² *True Politeness* was emphatic that under no circumstance was a lady to "boisterously salute" or, worse, shout out the name of a female acquaintance in public. To do so would compromise the privacy and decency of both ladies.⁵³ Such advice was seconded in the training that Miss Araminta Doolittle gave her students at the Rochester Female Academy. Her student Alice Hopkins later recalled, "What she wanted was to make us all over into high-bred, courteous, cultivated, truthful women of society, well-dressed and, above all, without eccentricities, trained never to do anything especially to attract attention."⁵⁴ The ideal lady was an invisible one.

Purposeful walking was critical to navigating the public landscape. As suggested by the terms *street walker* or *public woman,* such rules of conduct were needed to permit a genteel woman to be out in public.⁵⁵ Moralizing literature painted a picture of immoral women defined by the streets. The sentimental paean "Hymn for Female Penitents" published by a Troy authoress portrayed the fallen woman by her rambling habits: "Much hath she sinned—for many years /hath walked, by night, the city's street."⁵⁶ In the novella *Life in Rochester* Chumasero chose the suggestive name of "Eliza Streeter" for a naive girl ruined after becoming the lover of a man whom she had met on the street.⁵⁷ Phebe Davies, a seamstress of Syracuse, was sent to the Utica Insane Asylum by the orders of the county sheriff on the charge that she was a dangerous person to the community. Because her story is told by Davies herself, her transgressions are unclear, although it is indicative of her unconventional habits that when the authorities seized her she was found on the street "walking out for the benefit of my health."⁵⁸ By 1857 Syracuse authorities passed an ordinance directing that any woman "found loitering or strolling about the streets of the city, by day or night, without any lawful business" be fined ten to fifty dollars, the same fine charged to a convicted prostitute.⁵⁹ Etiquette guides stressed the importance of continuous movement while on the street. *True Politeness* specifically recommended bowing, not curtseying, when acknowledging acquaintances, because that protective form of greeting would not interrupt the flow and grace of the stroller's forward motion.⁶⁰ Appearances were everything; females had great access to public space but only if they kept moving.

The Syracuse teenager Augusta Rann adopted just such a continual "maneuvering" as her spatial strategy.⁶¹ She was an intensely peripatetic girl, whose walks took her not only through the Syracuse commercial dis-

male, blue-collar world.[132] A comparison of the bookseller's store with a lithographer's workshop shows that the retailer's need for transparency and visibility, architectural embellishment, and public access lessened in artisanal workshops, where women were rare (Figs. 36, 40).

Certainly, the white male was the main presence in the commercial district, as employer, employee, and customer, and he was well represented in the upper-story offices of the business blocks.[133] Even Chumasero's unflattering depiction of the greedy lawyer Daniel Grab upstairs, "in the back room of a dark, dusty, smoky law office, on Buffalo Street," tapped into the popular assumption that professional men kept their offices upstairs.[134] Augustus Strong was introduced to adult, male, white-collar, professional culture at his father's newspaper office in an upstairs business row. Working in the counting room of the Rochester *Daily Democrat*, the sixteen-year-old Strong learned the trades of double-entry bookkeeping, proofreading, taking telegraph reports, collecting bills, and running an office. The counting office, Strong explained, "was at that time the place of exchange for all Western New York." Equally important, he was immersed in the concentrated male commercial culture of the office. There, within what Strong called his "habitat," men discussed a wide range of topics: wheat crops, political elections, modern inventions, and religious philosophies.[135] These set-aside spaces became training grounds for the next generation of the mercantile city's economic, social, and political elites. The business publisher Freeman Hunt explained the spatial metamorphosis of the American young man's environs, "Just at the time that he was beginning to feel some interest in his studies, because he was beginning to understand them, he was cut short of any further instructions, and turned into the counting-house, to sigh for the green play-ground."[136] It could be a nearly round-the-clock spatial transformation. Edwin Scrantom recalled sad nights as a fourteen-year-old clerk after literally moving into his employer's newspaper office, where he had a "fellow apprentice and not a brother for a bed-fellow."[137]

By training young men for business, the ubiquitous mercantile college reiterated the gendered cast to the commercial district. Syracuse boasted four separate male business colleges, all renting quarters in the upper stories of the new business blocks, including the Italianate Pike Block on South Salina Street (Figs. 29, 41). Eastman's Model Mercantile College operated from the upper stories of Reynolds Arcade during the 1860s, from where it promised to "qualify young men of ordinary ability to take charge of a set of books in any establishment."[138]

The privacy of the upper stories also provided a protective privacy for the women and African Americans who plied their trades in the commercial district. For all the focus on inhibiting and regulating women on the street, the fact remains that women were participants in the commercial center not just as window shoppers or customers but also as shop clerks and entrepreneurs. In Hazen's *Panorama of Professions and Trades*, which admittedly was directed toward males, women showed up as workers in five trades—millinery and dressmaking, textile weaving, stitching in bookbinders, and clerking in confectioneries and jeweler shops.[139] Although Hazen should not be taken as the definitive statement on female employment, his attempt to present the typical types and settings of trades provides one normative view of the woman's place in the economic milieu of the city. Women's choices, it seemed, were either in the seclusions of the domestic arts or as window dressing burnishing life's little luxuries.

Self-employed businesswomen typically plied their trades in the less-expensive privacy of the upper floors, whose remoteness discouraged the random contact from impromptu window shoppers and the invasive glances of men. To be a street-level shop clerk was risking bourgeois sanction. In 1838 a New York newspaper decried that "the habit of employing girls in stores is becoming too fashionable. . . . It violates the natural mediety of the female character and strips it of the coy reserve which constitutes its chief loveliness." The "retirement of the domestic circle, and not the busy walk of commerce" was the "legitimate sphere of women," and one who transgressed against that position where nature placed her would loose her "caste" and endanger her virtue.[140]

In Chumasero's novella a simple immigrant girl, a stock character, stitched away in the classic "stock space" for respectable working women—the back room of the sputtering Mrs. Toddlecum's millinery.[141] Making his point of female seclusion, Hazen's millinery shop was shown without windows or doors (Fig. 39). Business directories carried several advertisements similar to Mrs. C. C. Van Every's millinery in Rochester indicating her shop was "upstairs" on State Street, where "a call is respectfully solicited."[142] In real life the Rochester seamstress Mary Farrell protected the respectability of herself and her State Street shop by enforcing its privacy. Located above a millinery and fancy goods store, which provided a thematically appropriate base, and situated on the second story, which offered respectable isolation, Farrell further protected her business by evicting loiterers. In throwing out a female browser, Farrell explained, "I said I would not allow them to sit in the shop and talk together; I don't allow any one to visit in my shop."[143]

The scanty record indicates that African Americans were supposed to be even more invisible than white women and that African-American storekeepers tended to avoid the publicity of the street level.[144] African-American businesses kept a low profile to avoid conflicts, even though many white residents of these upstate cities prided themselves on their abolitionist stance. Syracuse and Rochester were stations on the underground railroad, and black and white residents together developed refuges to thwart the southern bounty hunters. The upstairs Buffalo Street newspaper offices of the abolitionist Frederick Douglass held a secret compartment for hiding fugitives. In 1839 the Syracuse House hotel was the site of one dramatic slave rescue, in which African-American employees at the hotel initiated Harriet Powell's escape from her overnighting owners. In 1851 there was another prominent Syracuse rescue in which citizens broke into the jail to spirit off a fugitive slave captured by U.S. marshals.

Nonetheless, the racial tensions of daily life led many African-American entrepreneurs to minimize their public presence. In 1817 the fugitive slave Austin Seward opened a meat market in Rochester; a year later he built a two-story shop house in East Rochester and began a dry goods business. Seward set up several businesses in Rochester, always balancing the shopkeepers' need for a good location against the economic costs of such a move. Whereas residents typically looked at building as a sign of progress, as a black Seward was targeted for his ambition and had his shops torn down. Seward nonetheless continued in business and eventually invested in the most prominent business locations right near the Four Corners, first at the Rochester House and then on Buffalo Street across from the county courthouse. "We began to look up with hope and confidence in our final success," Seward remembered, but a suspicious fire around midnight thwarted his plans. "My store was on fire and a part of my goods in the street! . . . The building was greatly damaged and the goods they rescued nearly ruined. Now we were thrown out of business."[145] Being too prominent was a hazard, and moving upstairs or to cellar locations kept African Americans from being visible targets.

By midcentury there were several black-owned businesses in the central core business district of Buffalo Street, including a sail-manufacturing company in a loft, a doctor's office in an upper story of a business block, and several basement barbershops, including Bennett Jackson's barbershop under the Monroe Bank. Active abolitionists also worked downtown. Frederick Douglass operated his newspaper the *North Star* (in 1851 the name was changed to *Frederick Douglass' Newspaper*) from an upper story in the Talman Block on Buffalo Street catty-corner from the Four Corners. Harriet Jacobs similarly ran an antislavery reading room on an upper floor in the Talman Block.[146] Given the small percentage of blacks in the city, it seems likely that black businesses received white patronage. Nonetheless, none of these businesses opened directly onto the street, a move that provided the safety of privacy as well as less expensive quarters.

Personally, African-American leaders found a mixed reception within the social landscape of the commercial district. Douglass lectured weekly one season at the popular Corinthian Hall owned by William Reynolds, "who, though he was not an abolitionist, was a lover of fair play and was willing to allow me to be heard. If in those lectures I did not make abolitionists, I did succeed in making tolerant the moral atmosphere in Rochester."[147] It was at best an ambiguous toleration. Invited to a printer's reception in the Irving House hotel in 1848, Douglass was blocked at the dining room door by the hotel keeper, who claimed it was "a violation of the rules of the society for colored people to associate with whites."[148] An awkward vote by the assembled printers gave Douglass the majority, and he was invited in. "It was a painful, as well triumphant hour."[149]

Lower-class blacks faced greater hostility, even as they contributed to the economic vitality of the city. In 1842 the Rochester shop clerk Lindley Gould snickered about a trick he had played on an illiterate black man Andrew Wilbur, whom he called "Black Jack." Wilbur came into his store seeking a written order and unwittingly left not with the fifty-cent order to take on to the next errand but, rather, an "order requesting Mr. Squires to kick the nigger out of the shop."[150] The white merchants who tried to regulate the commercial, visual, and social tenor of the downtown were not a monolithic class when it came to race relations, yet it seems possible that the sequestered black landscape of their fellow entrepreneurs suited even the liberal elites, who were looking literally and figuratively to whitewash and smoothen any rough edges of the cityscape.

Top Floor

Assembly halls inserted into the top floors of the business blocks reiterated the pattern of privatized public space which characterized the commercial city. Landlords faced particular challenges in renting out the inconvenient rooms at the very top of the stairs. Instead of carving out low-rent offices, they often configured the top floors into special-purpose open halls appropriate for large groups whose necessity for meeting space outweighed the inconvenience of the ascent. The landlord's economic decision carried bourgeois social overtones. Fraternal, religious, political, temperance, and civic groups rented space in these large halls and thus in theory represented the kind of responsible citizenry bent on self-improvement and civic participation that would gentrify the new city's physical and cultural landscape. Once again, commercially conditioned social space worked toward a bourgeois vision of the good city.

Halls developed particular reputations and had the power to cast favor on the events as well. Rochester's Monroe Hall specialized in temperance societies, with five different temperance societies regularly scheduled for evening meetings and the ladies' Washington Society electing daytime meetings. The Rochester Odd Fellows Hall similarly hosted five different groups meeting on five different evenings. The confluence of available rooms, convenient location, proximity to work, public separation from the private domestic hearth, and a business atmosphere made the commercial center an attractive locale for social organizations. The temperance-focused nature of many of the groups lent a respectable air to sociable evening outings in a city where taverns offered the more common nighttime diversions. The time pulse of these spaces also contributed to the social gentrification of the commercial district. In the evening, when business shut down, the halls opened up, drawing a crowd into the streets and buildings. Also used for political conventions, student examinations, mechanics fairs, agricultural society banquets, literary society meetings, halls became part of the public sphere of civic citizenship.

Yates, Top to Bottom

During the second half of the nineteenth century the increased capitalization of manufacturing and commerce resulted in vertically integrated businesses within single, large business blocks. Although it was architecturally similar on the outside to the multi-tenanted 1850 Pike Block, the 1865 Yates Block was organized internally as a single-business enterprise (Figs. 29, 42). It encapsulated the new trends in architecture, manufacturing, and commerce. Having worked in Rochester and Utica, Alonzo C. Yates opened up a clothing store in a rented Syracuse shop in 1851, expanded into the adjoining store in 1856, and purchased the entire building in 1857. Yates made his fortune in men's ready-made clothing, a business that ballooned during the Civil War, when uniforms were needed. Between 1863 and 1865 Yates constructed a new building specifically suited for manufacturing and selling men's ready-made clothing. Located on

North Salina Street across from the Empire Block, the building fit into the evolution of dry goods row from a place selling the materials for clothing into a place selling the completed item. Both in style and internal organization the Yates Block summarized the urban, architectural, and sociospatial ordering of the era.[151]

The five-story, fifty-foot wide, Italianate exterior received the expected platitudes due "that ornament to our city." Yates was locally famous for his grandiose architectural tastes, having become "fascinated with the castle style of architecture during his first tour of Europe" and later purchasing an 1852 Gothic Revival home designed by the prominent architect James Renwick. But the inside of his business block also garnered attention: "No one who has not visited this great palace of art and industry can form the faintest idea of its vast proportions, its clerks, salesmen and operatives. One can scarcely conceive where a market can be found for such immense quantities of clothing of every kind, quality, style and fashion; from the rough garments of boatman and dither to the elegant and costly apparel of the nabob and dandy." Cast-iron columns replaced partitions between the double-wide shop floors, making one great room but also "ornamenting and relieving the appearance" of the interiors. The interior sorted activities and people by floor, yet all were linked vertically—visually by illuminating skylights and light wells, and audibly by speaking tubes (Figs. 43–44). The first floor was dedicated to the retail trade, where under the globe lamps were tables piled high with heaps of men's clothing; a few female shoppers are depicted in images from the era, but the clerks and clientele were typically male. Environmental expectations had risen even for the men now, and in the rear of the floor was "the cozy office and countingroom, and a real boudoir in appearance."[152] The second floor was the general wholesale room for piece goods and trimmings. Women were again outnumbered by the male jobbers, who made the bulk of the purchases, although the shawled and headdressed females were all accompanied by escorts, whose own dress and stovepipe hats similarly encoded them as members of the bourgeois class.[153] The third floor was divided into a cutting room with men and boy operatives and a salesroom for coarse garments catering to a male clientele. The fourth floor was the great wholesale room, where Yates joked that he sold clothes "by the thousand, cord or ton."[154] Instead of an open sales floor, low dividers separated the stocks. Women would have few reasons to enter a wholesaling space, and the single one depicted in the illustration seems to be accompanying the top-hatted gentleman. The fifth floor held the manufacturing room, where female operatives were hidden from public view as they stitched in neat rows under the watchful eye of male supervisors.

The Yates Block manifested multiple aspects of the sorted city. The building participated in the specialization of dry goods row. The exterior was part of the architectural refinement of the era. The interior progressively refined the sorting by activity and gender and did so in an orderly way: "And yet in this great establishment there is such a perfect system, such exact attention to the regulations and such a prompt obedience to the established order of business, that there is no noise, no confusion, no apparent haste, no unpleasant jostling or interference; but everything moves on with the regularity of a great machine, and with an ease and quietue that would not disturb a parlor."[155] Lastly, the richest man in Syracuse, Yates himself was heralded by the business community as embodying the urban patron through his construction of an ornament to the city, employment of locals, and the example of his personal habits. "It is in this way that prosperous business men are the real benefactors of the community."[156] A relief of a bespectacled man, Yates perhaps, crowned the parapet of his building, beholding the city below. Metal store tokens imprinted with image and slogan of "the old man with specs" doubled as advertisements and small change, further circulating Yates through the city.[157] This was a merchant who claimed public space in a manner larger than life.

SIX

The Reynolds Arcade and Athenaeum

Reynolds Arcade in Rochester epitomized the merchant's sorted cityscape of the nineteenth century. Completed in 1829 and remodeled at midcentury, the Arcade was a landmark piece of urban promotion that continued to be an important touchstone in the social and aesthetic ordering and physical renewal of the maturing city. One newspaper described the Arcade as "the chief of landmarks, the most beaten thoroughfare, the very vortex of the city's new and great place of congregation of the people at all times."[1]

Geographically, the Arcade reflected the anticipated and actual rise in land values and the expansion of the commercial district. Functionally, it promoted overlapping economic relationships, creating webs of commercial exchange among the landlord Reynolds; his shopkeeper tenants; the eastern manufacturers, who produced most of the goods for sale in the Arcade; members of the local rising professional class who offered their services from Arcade offices; the customers who purchased their goods and services; and the public who depended on the intelligence gleaned from the mail delivered to the Arcade post office. Architecturally, the Arcade helped to define the commercial district and then preserve it. Culturally, it was an eloquent symbol of Rochester's social and architectural refinement. The rare building type distinguished Rochester, giving it a particular panache among all American cities. The Arcade embodied the aspirations of the merchant class, whose members sought an efficient, profitable, beautiful commercial experience whose very space, products, and image served their business and social needs. Moreover, the way in which the ensemble of Arcade buildings cast out peddlers and wastrels eloquently testified to the Arcade's powers of social gentrification.

The Architecture of Commercial Competition

Reynolds Arcade established, maximized, and refined the identity of the Four Corners as Rochester's premier commercial district. Both its construction and remodeling showed the response of the Reynolds family to two parallel commercial conditions within the city. One was simply the economic opportunity to maximize the productivity of their centrally located double lots by exploiting the interior of the block. The other was to draw business specifically to this site and away from rival sites (Fig. 45). By the time Abelard Reynolds built his arcade, the commercial district was already pressing beyond its Four Corners location. Even the Genesee River was unable to stop the commercial tide, and merchants built their shops directly on the roadbed of the Buffalo Street bridge.[2] In 1828 the east bank developer Elisha Johnson had built a massive five-story, 130–room hive known as the Globe Buildings at the other end of the Buffalo Street bridge on what was called Main Street. The expanding commercial vigor that fueled the economy, however, threatened the old center.

Abelard Reynolds's allegiance to redeveloping his double lots on Buffalo Street was both pragmatic and political. The east bank settlement of Brighton township developed by Elisha Johnson had been officially annexed to west bank Rochester in 1823, and the Buffalo Street bridge linked both banks in a continuous line of commerce. Nonetheless, political and commercial rivalries between the two sides remained. Having settled in Rochester in 1813, the pioneering Reynolds developed strong and lasting ties to the west bank and to Nathaniel Rochester, who had helped secure Reynolds's appointment as postmaster, appointed him to the board of directors of the Bank of Rochester, and supported his successful candidacy for the New York legislature.[3] Johnson had been in the area nearly as long but as an investor in rival settlements on the east bank. He had been sued by Nathaniel Rochester regarding the construction of the east bank mill dam and subsequent flooding of Rochester in 1817, and, as a Democrat, Johnson had continued to oppose Nathaniel Rochester's Republican maneuvering in local and state politics. Thus, Johnson's expansive Globe Buildings on Main Street not only threatened the commercial dominance of the Four Corners shops, but it also personally challenged the economic and social hegemony of the Four Corners founders. The fact that Vandewater's 1831 tourist guide recommended seeing this complex as part of any tour of Rochester confirmed the east bank's rise.[4]

Reynolds immediately responded to Johnson by constructing his unique Arcade. He relocated his wood-framed tavern, saddlery, and post office building to the rear of the lot to make room for the new Arcade. As originally built, the four-and-a-half-story Arcade ran back fifty-six feet and presented six separate storefronts running ninety-nine feet along Buffalo Street. A wing connected the central open hall of this front Arcade to a rear Arcade measuring sixty by ninety feet. Behind this lay the old frame tavern. The roof was crowned by a distinctive public observatory "in the form of Chinese pagoda" that towered ninety feet above the street, making the Arcade both a sight to see as well as a place from which to see the sights (Figs. 14, 46).[5] It is unclear to what degree Reynolds was aware of the arcades being erected in New York City, Philadelphia, Providence, and Stonington, Connecticut, but his motivations and strategy echo those of Cyrus Butler in Providence. That arcade, too, was built as a response to changing real estate interests. The Rochester Arcade, however, was built to reaffirm the centrality of the original commercial core, not to expand it.[6]

Increasing land values in the Four Corners prompted Reynolds to intensify his utilization of the Arcade's double lot. After only a decade of use, the Arcade underwent a series of remodelings and tinkering. In 1838 Reynolds again relocated his frame tavern, this time to the other side of Bugle Alley in the back. In its place he constructed another connected building dedicated solely to the post office. In 1842 Reynolds unified all three Arcade buildings into a single Arcade comprising eighty-six rooms subdivided into forty-two rental properties, the whole edifice stretching continuously from Buffalo Street to Bugle Alley (later known as "Works Street"). The post office was located in the rear northwest corner of the Arcade until 1859, when it moved across the hall to the northeast corner; a lateral addition to the east in 1862 provided yet more space for the postal and telegraph offices. Reynolds's decision to extend the number of shops, continually relocating the tavern building and the institution of the post office, and then by extending the Arcade itself, was probably motivated by the increased rents he could get for the shops within a desirable central location. The recessed location of the post office and tavern served as magnets pulling people through the Arcade and thus maintained the flow of traffic through the passage.[7]

The architectural refinement of the commercial district proved to be a never-ending process even for such celebrities as the Reynolds Arcade. The fashioned and refashioned Arcade buildings were gifts to the street, acts of architectural patronage on the part of Reynolds (Fig. 46). At the same time, they were also rhetorical slaps to business competitors. Looking back in 1887, the Rochester *Herald* recalled the architectural climate at midcentury:

"The owners of central property discovered that if they would retain the advantage they had so long enjoyed they must look to something beside locality, and that they must erect larger and better buildings to hold their advantage."[8] Commerce had been stretching not only along Buffalo Street but also up State Street. As part of the architectural refinement of the commercial district, other merchants were investing in newer and more fashionable business blocks. A rival dry goods firm had moved off the Buffalo Street bridge to State Street, and in 1846 Abelard's son William was drawn into the architectural fray, galled by the competition's "elegant Store," which was "very beautifully & tastefully arranged, and finished—but the principal point of attraction is the front of plate glass, which is very rich."[9] The owner of the Burns Building next door had also been making improvements, adding plate glass windows between new cast-iron columns and adding another story for a new Masonic Hall. William Reynolds worried that "it makes our building look flat" and threatened to eclipse the Arcade's rooftop observatory.[10]

Pride was only part of the problem. Watching the emergence of a luxury dry goods row on State Street, William Reynolds warned his father, "I find public sentiment is very strong against Buffalo Street for Dry Goods & unless Something is done Soon to render Buffalo St Stores more attractive, we shall have to abandon that description of business & fit up the Stores for some other business."[11] This was a challenge that the Reynolds could not afford to decline.

Intent on changing the Arcade's fortunes, father and son began to implement a series of architectural interventions intended to restore the building's original stature.[12] Aesthetic awakenings required architectural patrons to look beyond the local area for ideas and materials. Just as Henry Dillaye had traveled to the metropolises back east in search of architectural ideas for his Washington Block in Syracuse, the Reynolds family inspected New York City buildings. William specifically instructed his father to inspect the glass and shutters in the city and to take a look at the New York showcases, including the fabled Marble Palace of A. T. Stewart on Broadway.[13] In this climate of visual refinements Abelard and William Reynolds finally turned to an architect for assistance. The local architect D. C. McCallum sketched out his plans and proposed changes typical of the period: removing several piers and replacing them with cast-iron columns and reconfiguring the facade into a series of recessed entrances filled with plate glass windows. William Reynolds supported the idea of the renovation and defended its expense to his father: "It would be Splendid improvement for the Hall &

would make the 2 stores at the entrance very showy & desirable. . . . The whole improvement . . . will cost 7 or $800—which I am aware is a great deal of money to spend—but I do believe it will be a good investment, in the influence it will have in enabling us to rent the rooms in the Hall—without something to give a new impulse to things, We Cannot Sustain the rents in the Hall."[14] Fashion, he was convinced, could "give a new impulse to things."[15] The remodeled facade fit in perfectly with the prevailing taste for order and uniformity. Rows of plate glass, open-fronted shops marched across the facade. The observation tower was an architectural exclamation point to the Arcade's fashion statement.

Abelard and William conceived of the remodeled Arcade as the setting for a full-fledged genteel tableau (Fig. 47). The public entered the Buffalo Street front through a narrow lobby graced by scenic paintings of the Genesee and Niagara Falls. The Arcade hall then opened up as a glass-covered pedestrian street running the depth of the block and rising fifty feet high. The architectural details of cast-iron shop fronts and balconies created an airy impression under the skylit hall and decorated lobby. The shops were politely retail or respectably professional; no butchers, hardware merchants, or artisans invaded such genteel settings. Before it relocated to the Athenaeum, the Mechanics Association reading room lent its cultured touch with inspirational busts of national and classical heroes, including DeWitt Clinton, Benjamin Franklin, Cicero, and Homer, interspersed among an ever expanding collection of books.[16] With its combination of commercial shops at street level and the self-help organization of the Mechanics' reading room on the second floor, the Arcade picked up on the idea of social opportunity through economic achievement, a perfect virtue within a mercantile city.

William Reynolds's midcentury plans to reinvigorate the Arcade Building mushroomed as the architectural standards and financial stakes escalated. The $800 estimate William had floated by his father in 1846 had grown to $12,000 by 1848, and even William admitted, "I am building much larger & more expensive than I designed when I made the loan."[17] The mounting debts, however, were not solely caused by the Arcade.

In addition to improving the architectural amenities of the Arcade Building, in 1849 William Reynolds built the completely new, freestanding Athenaeum deep in the interior of the block, across the back alley from the Arcade (Fig. 45). Not coincidentally, one of the primary ways to access this new building was through the passage of the Arcade. Reynolds was convinced that the Athenaeum would draw more traffic through the Arcade, reviving the older building's fortunes while making new ones.[18] The key to either building's success would be to make the Athenaeum magnetic enough to attract the public. Embedded deep within the block, the first-floor shops opened up new retailing and shopping opportunities in the Four Corners area. The Mechanics Association on the second floor was a stand-alone business that brought its own clientele. Its library and meeting rooms enjoyed an automatic audience of members. The third-floor Corinthian Hall was elaborately designed to become the premier assembly hall in the city, drawing daytime exhibits and nighttime entertainments.

Functionalism made these spaces attractive to Rochesterians, but so, too, did beauty. Attention to appearances was now expected in architectural patronage, and Reynolds assured his public that the new edifice would hold "one of the most splendid halls of the kind in western New York. . . . The whole will be finished in the latest style, and made substantial and lasting."[19] The local architect Henry Searll designed a blocky three-story brick edifice, embellished with Italianate details at the windows and cornice, the whole dominated by a yawning aedicule framing the imposing, two-story entryway, whose classical entablature was inscribed "Athenaeum" and "Mechanics Association" (Fig. 49).[20] William Reynolds's words showed how deeply commercial ideals had co-opted the civic. He proudly declared that the "lofty Grecian entrance" of the Atheneum would bestow upon the commercial building the dignified "appearance of a public edifice" and promised that it "will be an ornament to that part of the city."[21] The coup de grâce was the Corinthian Hall on the top floor (Fig. 50). Known for its architecture, the assembly hall was luxuriously embellished with two Corinthian columns flanking the stage and Corinthian pilasters alternating with twenty-eight windows standing sixteen feet tall. The lofty ceilings were twenty-six feet high, and the capacious room, with its much-heralded acoustics, was capable of holding hundreds of attendees, reaching full capacity with an estimated sixteen hundred guests at one overflowing event.[22]

The Reynoldses' responses to commercial competition had stimulated both an economic and architectural renewal at the Four Corners. Edwin Scrantom marveled over the architectural metamorphosis of the alley that accessed both the rear of the Arcade and the front of the Athenaeum. "The same transforming element has been at work on Mill street and points adjacent. Where, in the recollection of the writer, were formerly a few cabins or lines of barns and out-houses, now stand in majestic column, long rows of splendid brick and stone edifices, and modern built mercantile houses for wholesale trade, and the place is no longer 'tin-pot-alley,' or 'bugle alley,' or 'skunk lane,' as it once was, but a finished street with granite side-walks, and buildings and streets in the royal finish and perfection, that are beyond any former time" (Fig. 51).[23] Commenting on the new buildings that faced the improved alley, a local newspaper crowed that Rochester was gaining on the urban leaders: "Works street will be able to boast of as much fine architecture as any other street in the city, not only of its size, but of any size."[24]

Urban Promotion

The new nineteenth-century mercantile cities needed to develop extra-local reputations in order to attract human and economic investment. The Reynolds family acted as classic city boosters when they constructed the ever-expanding Arcade. The building made an ambitious statement of urban conditions and urbane culture. The Reynolds Arcade bolstered the business interests of the Reynolds family, their Four Corners neighborhood, and the wider city. The Arcade was a place-making piece of architecture which helped to put Rochester on the map. A showy hotel garnered attention, but the rarity of an arcade put Rochester in a special class of American and European cities. Residents took particular pride in their Arcade, pointing it out to visitors and immortalizing it in local poetry. Penned one hotelier in his distinctive Scottish brogue: "We hae a splendid, bauld *Arcade,* Which leaves all ithers in the shade."[25] They even produced musical compositions such as "The Rochester Arcade Quick Step," with the building emblazoned on the cover of the sheet music (Fig. 46).[26] The sophistication of the Arcade and the gentility of its shops encouraged metropolitan comparisons with mixed results. Whereas some Europeans found Rochester and its arcade quite provincial, others were impressed, such as Frederika von Bremer, who commented on the "handsome, well-lighted room in a large, covered arcade, in which were ornamental shops like those arched bazaar-arcades in Paris and London."[27] Homespun Americans were more easily impressed. Visiting the Arcade in 1831, a Long Island shoemaker marveled: "the stranger standing in the midst of this place can hardly believe that only 20 short years since the spot on which this town stands was a howling wilderness. This place has all the appearance of splendor and fashion of our sea board populas [sic] towns."[28] In 1856 the remodeled glass and iron interior prompted the honorific title of Rochester's "permanent crystal palace," a tribute surpassing the temporary exhibition architecture of London's Crystal Palace of 1851 as well as New York City's 1853 version.[29]

To be fair, the Arcade paled in comparison to the arcades of Europe, and it was also a plainer version than its few American counterparts. The Arcade's 1829 brick facade and awkward cupola contrasted poorly with the 1828 arcade in Providence, Rhode Island, with its sophisticated classical detailing and interpenetrating granite facade. But the Arcade was not competing with the arcade in Providence or any other east coast established city; it was competing with East Rochester and other nearby upstarts. In this capacity it was quite successful. Like it or not, Reynolds Arcade was an extraordinary piece of architecture that was literally remarkable. It helped establish the urban credibility of Rochester and the New York interior. As the resident Jenny Marsh Parker proudly explained, "it stamped our individuality when we were hardly expected to have individuality."[30]

In keeping with the mercantilist city-building endeavor, the Arcade fostered the business of business in a larger geo-economic sense, not simply by renting quarters to retailers and professionals but, more broadly, by providing the space for institutions that forged links along the mercantile chain of cities and even across the nation. As with any shop, the sale of eastern-imported goods sent profits across the state, and the sale of locally made necessities stimulated regional production. For that reason alone the Arcade would have been economically important. But the Arcade held more than shops. William Reynolds described the Arcade as a "sort of Merchants exchange for this city."[31] Holding the post office made the Arcade one of the more important spaces of residents' daily routine. Being in the information loop was critical in gaining commercial advantages. Post offices were critical to a city's development. Nathaniel Rochester had pushed to get Rochester a designated postal stop that would make it the central source for commercial intelligence, and Syracuse had even changed its name from Corinth in order to receive a post office. Noting the benefit of the post office to Rochester's economic and cultural development, an 1836 article in the nationally distributed *American Magazine of Useful Knowledge* remarked, "the annual income of its post-office, which is a good test both of its literary taste and commercial prosperity, is over $14,000."[32] The post office was also a marketing tool for Arcade businesses. The recessed location of the post office drew people into the Arcade and along past the shops. During the 1840s the new telegraph office joined the post office in the Arcade, making it the magnetic hub of commercial intelligence for local and regional business interests. The Arcade contained professional offices upstairs, including for a few years those of the prominent magazine the *Genesee*

Farmer. The magazine's location in Rochester not only symbolized Rochester's connections to its hinterland but also gave merchants the easiest access to hinterland strategies that might, in turn, influence their own merchandising, milling, or manufacturing decisions. The magazine also placed Rochester squarely in the middle of a national readership interested in agrarian, rural, and market improvements, all key concerns in the mercantile expansion of cities and market on the eastern frontier.

The Arcade therefore provided not just goods and services but also communication, linking people to events in the world outside. Access to current information was vital in any business dealings. Cities with the best information links—those having access to newspapers, post offices, telegraph offices, and business travelers on active transportation lines—became the major economic centers. The expansion of "intelligence flows" resulted in the extension of the wholesaler's trade area and thus commercial profits. The benefits of the rapid circulation of economic information accrued most quickly and conveniently to businesses close to those circulation points. The post office, telegraph office, canal offices, and eventually train stations were never far from one another, nor were they far from the merchants who were accordingly drawn to and contributed to the centrality of the commercial district as the communications site. The major economic actors therefore clustered near these relay points. By 1863 the Arcade held not only the post office and Western Union Telegraph office but also four insurance agencies, three real estate brokers, one commercial information agency, one banking office, and the internal revenue collector's office.[33] Thus, the Arcade was a clearinghouse as well as a market house for mercantilist opportunities. Information, ideas, lines of credit, and cash flowed through the Arcade's halls.

Commercially Conditioned Public Space

Both the interior space and exterior presence of the collective Arcade buildings acted as tools to promote the kind of businesslike and genteel behavior advocated by the bourgeois city leaders. By midcentury the local newspaper could report that, among the "concentration of so many places of business," the Arcade Building was the preeminent "place of public resort" from dawn to late at night, thronged with people "giving an air of cheerful bustle and activity."[34] The building was an exemplar of the city's commercially conditioned public space, not only within its walls but also without.

Each improvement swelled the Arcade's business. A more critical appraisal might have noted that the primary clientele, however, was the city's bourgeoisie, who benefited from the creation of genteel spaces.

The Arcade enhanced the personal display of gentility by both tenants and customers and thus was a space that was not only good for business but also good for enacting the class-based ritual of the promenade. An 1851 illustration depicted the genteel spectacle of frock-coated gentlemen with top hats and walking sticks and demurely-dressed ladies in shawls and bonnets accompanied by companions or children (Fig. 47). Some of the customers are peering into shop windows; others have stopped their stroll for the simple pleasures of conversation. From the mezzanine professionals and their clients enjoy the prospect below as well as their own convivial display. The social and spatial tableau combined self-consciousness with niceties and reinforced the bourgeois code enactment of social order. "The Arcade was a common rendezvous where busy citizens said 'good morning,' and those from the country round about, who meet less frequently, 'how are you?' and we believe it contributed not a little, though of course in a most incidental way, to that general acquaintance and good understanding which is so important in every well ordered community."[35]

Functionally, architecturally, and rhetorically, the new Athenaeum explicitly commingled the idea of civic and commercial space. Cloaked in both the nomenclature and style of classical humanism, the Athenaeum and Corinthian Hall invoked a sense of civic purpose. The very type of the institution—an athenaeum—romantically harkened back to the idea of a democratic civic space in which ideas and learning were shared. And the naming and outfitting of the public assembly room—the Corinthian Hall—redoubled the classical allusion to ancient Greece. Because ancient Corinth was famed as a trade center, it is also possible that the nineteenth-century literati considered the name Corinthian particularly appropriate for a landmark gracing a mercantile city.[36] William Reynolds pushed the idea of the Athenaeum as a civic edifice, with himself as the architectural (and hence civic) benefactor. Merchants had once again used architectural patronage to reiterate their cultural and economic authority in the city while simultaneously creating the kinds of settings which suited their own personal and class interests. The mingling of civic and commercial connotations in the Athenaeum never threatened the integrity of the actual civic district, composed of the courthouse and what had grown to be four churches on the Buffalo and Fitzhugh streets axis, but, rather, perpetuated the early principle that what was good for commerce was good for the *civitas*.

Even with the presence of a private watchman, however, the tableau crafted by the assembled Arcade buildings presented a compromised gentility. The Buffalo Street Arcade provided only a partial solution to the physical and social unpleasantries of urban life. The skylit hall protected the public from foul weather and mucky streets, but anybody could enter the Arcade. Rambunctious boys had to be reigned in from spitting on the mural. The post office attracted all types of people. Both the front and back entrances to the Arcade were bracketed by the very people the arcading public preferred to avoid: the blind organ grinder camped out by the front doors and the old woman selling matches out back.[37] In addition, the Reynoldses' attempts to hold onto the dry goods business had obviously failed. There was only one merchant tailor dealing in cloth by the time an 1863 directory of the Arcade was published, and he operated in the side wing of an upper floor.[38]

The 1863 directory provides a useful glimpse into the spatial sorting within the Arcade. Six shops fronted Buffalo Street without connecting to the Arcade (Fig. 48). These shops were all part of the commercial mix on the street: two tailors, a bookstore, a druggist, a banking and insurance office, and a shoe store. Flanking the lobby entrance, the public was greeted by Dewey's expansive bookstore and a jeweler, creating a suitably genteel entree for the bourgeoisie, including females. But, in fact, the shops deeper into the hall were oriented to men, with insurance, real estate, a cigar store, and the eating, drinking, and smoking rooms of the Arcade House, all tapping into the clientele pulled in by the post and telegraph offices. A "hair cutter" (as opposed to barber) advertised that he did ladies, gentlemen, and children's hair cutting, but it seems doubtful that ladies actually came to the Arcade shop. The second story similarly held professional offices that employed and catered largely to men. In addition to the Reynoldses' business office there were at least ten attorney offices, three real estate offices, the internal revenue collector's office, the customhouse clerk's office, plus a dentist, barber, and architect. Indicative of their status in the city, the middle-class male had earned his rarefied space.

The upper floors assumed a decidedly artistic cast, containing at least fifteen art studios, including artists, sign painters and wood grainers, an engraver, a drawing teacher, landscape painters, portrait painters, a sculptor, several daguerreotype studios, and even a designer of artificial limbs whose artistry resembled "Nature's own handiwork."[39] As an early Bohemian quarter, the artist colony in the upper reaches of the Arcade catered to the odd mix of the elite and the down and out. Here mixed a cultured public whose members sought artwork or art lessons, members of the general public who were drawn to the plush parlors and affordable prices of the daguerreotype studios, and the jarring contrast of mutilated mill and factory operatives needing prosthetics. The third-story remove of the locations perhaps also gave these cheaper quarters genteel associations of privacy. Whether taking drawing lessons from Miss E. L. Smith or posing for a portrait, the female customer would not have been sitting on display. The daguerrean studios explicitly advertised special accommodations for ladies, including dressing rooms and salons. The architectural setting, however, retained its powerful associations of gentility, preserving it as a place of popular resort for a variety of residents. Although inconsistent, the overall tenor was one of unique, rarefied purpose that gave the Arcade an elevated identity.

At street level the tenants at the Athenaeum, including a confectioner, were similar to those of the Arcade. But the clients became more selectively clubby on the upper stories. The reading room was open to the male members of the sponsoring institution and to all ladies and gentlemen; Alcesta Huntington came here to read while waiting for her ride home.[40] The Corinthian Hall was the venue for numerous events whose only occasional exclusionary feature was charging admission. But the alley it fronted was hardly a genteel entry.

Analyzed at the neighborhood scale, rather than just the building scale, the Arcade and Athenaeum projects were part of a larger social reclamation project that favored the interlaced bourgeois and mercantile cultures dominating the new cities. Explicitly, the Arcade was about profit maximization, reclaiming the Four Corners as the preeminent commercial location and creating a genteel setting for social and commercial relations. Implicitly, the Arcade was about literally removing and figuratively marginalizing the people and activities that challenged the economic and social refinement of the building and its larger neighborhood. The Arcade buildings not only cut *into* the interior block but also cut *out* what was there—mucky alleys littered with ramshackle buildings and a cast of social undesirables. Redevelopment transformed nefarious Bugle Alley behind the Arcade, first into the productive Works Street and then into the fashionable Exchange Place that fronted the Athenaeum. The very renaming of the interior alley traced the upgrading of the street's contribution to the city's image.

Remembered as an "abandoned quagmire, the common garbage receptacle of the neighborhood, unpaved and unlighted," Bugle Alley was a problematic rear entrance to the Arcade and an even less auspicious front entrance to the Athenaeum.[41] The longtime Buffalo Street businessman Edwin Scrantom recalled the general tenor of the back alley: in addition to housing quarrelsome and inebriated Irish, the "back of the Arcade was the great place in early days for Indian 'drunks.'" He proceeded, "Unlike today, that place was a great hiding place, and one of quiet; until, from the many mixtures of depraved human nature there, it got the name of 'tin-pot-alley,' and afterwards 'bugle alley.'"[42] Even before the Arcade was constructed, the citizens of Rochester, including Abelard Reynolds, had formed a public watch to dispel the disturbing increase in vice and crime in the city. Failing as a private organization, they supported the establishment of a police force to maintain order.[43] Although crime was a concern, it appears that vice was the greater problem in the secluded back alley. The police blotter in 1837 recorded the following incidents:

> July 29, Patrick McKann found on Work St. Drunk & Asleep.

> August 26, Sarah Owen (alias) Sarah Collins found on Works Street some Drunk at 1/2 past 10 at night . . . police charge disorderly & vagrancy Com. Prostitute.

> Sept 17, James Mix found back of Arcade at Mr. Thomas Beards Deranged put in W[atch] House before police charged Insane (Delivered over to Poor Master).[44]

Bugle Alley, in short, was a social problem for the likes of the founding fathers and city leaders, including the Reynoldses, the Scrantoms, and the Rochesters. The following year Abelard Reynolds began his first remodeling of his ten-year-old Arcade.

By cutting into the interior of the alley, erecting showy pieces of architecture, and drawing a continual throng of people through the interior of the block, the Reynoldses' construction projects literally exposed the litter of activities and personalities which had challenged the bourgeois merchant class's ideas about both the economic efficiency and social gentility appropriate to a commercial district. Architecture and urban planning exposed and refined these hidden urban spaces and their occupants. The Reynoldses' constant building projects maintained the morphological, architectural, and social pressure on the secluded back alleys. Visibility and order went hand in hand. The commercial redevelopment of the alley tried to push these unwanted denizens out of the gentrifying commercial district by exposing their hidden ways to public scrutiny. It was a distinctly bourgeois response to the deeper structural problems of high real estate costs, uneven development, and the absence of social services. Where these denizens went, let alone the causes of their derelict behavior, was not the primary concern. Rather, the impetus was on establishing an economically productive, visually and socially sorted, ordered commercial district.

The 1849 addition of the Athenaeum increased the pressure by routing bourgeois shoppers, reading room members, and the general public attending fairs and events through the alley. What daylight and shop fronts could not achieve, perhaps a disquieting bourgeois throng could. Evening events in the Corinthian Hall were particularly helpful in maintaining public oversight of the secluded and darkened backstreets. Known for its edifying entertainments, national luminaries such as Daniel Webster, Horace Greeley, Ralph Waldo Emerson, Oliver Wendell Holmes, and Louis Agassiz delivered public lectures there (Fig. 50). The right setting, which included high-style interior design, such as the Corinthian Hall's columned and pilastered space, helped improve the tenor of any event. Even the views of controversial speakers such as the women's rights advocate Victoria Woodhull, the abolitionist Frederick Douglass, and the phrenologist Orson Squire Fowler, who spoke about marriage and "sexuality and creative economies," became polite subjects for consideration within such architecturally elevated surroundings. It similarly sanctioned amusement by providing elegant conditions for such eye-opening performances as those by the actress Fanny Kemble and the dancer Lola Montez. The ladies of the Rochester Charitable Society, who had a generation earlier refused to sanction theatrical events by accepting any proceeds, changed their tune when offered a share in the receipts of the "Swedish Nightingale" Jenny Lind's recital in the Corinthian Hall.[45] When the newspaper reported that the Arcade buildings were a "place of public resort" from dawn to late at night, thronged with people "giving an air of cheerful bustle and activity," it was acknowledging the social hygiene of the setting as much as its popularity.[46]

Citizen Reynolds

The linkages between the merchant and the mercantile city which had implicitly guided the economic and cultural colonization of the interior found full voice in the expressions of mourning at the 1872 death of William Reynolds, or, as one memorialist called him, "Citizen Reynolds."[47] William Reynolds became the exemplar of the merchant as public citizen. The numerous memorials and eulogies made in his honor all emphasized that

Reynolds was "an eminently public-spirited man," and "remarkable for his enterprising public spirit."[48] Reynolds's municipal leadership led him to serve on the City Common Council, Trustees of the Rochester Savings Bank, Trustees of the University of Rochester, the Managers of the House of Refuge, and the Board of the Athenaeum and Mechanics Association.

Yet it was not just his social responsibility that earned him accolades. It was also his fulfillment of the idea of merchant as the patron of the city. As if to fulfill the challenge raised by architectural and cultural critics such as James Jackson Jarves and David Buel, the Reynoldses had used architecture to build their own reputations and that of their city. The father, Abelard, was praised for the construction of the landmark Arcade building, and his son William was credited with continuing the legacy. "His history and that of this city are closely allied," penned the mourning Arcade tenants; "no man has more completely identified himself with the growth and vigorous prosperity of Rochester during the several stages of her history than he."[49] The local newspapers equated Reynolds with the city: "As regards the public and business life of the deceased, it is the history of the prosperity of Rochester," and then it proceeded directly into a discussion of the Arcade and Corinthian Hall as proof.[50]

As patrons of architecture, the Reynoldses had bestowed gifts to the street and city. The City Common Council specifically paid tribute that William's "laudable pride and ambition ever was to make this city beautiful and attractive."[51] The *Rochester Union and Advertiser* specifically pointed out how the "beautiful" Corinthian Hall redeemed "one of the most disagreeable spots in the city."[52]

Memorialists ascribed a certain morality to William Reynolds's building decisions: "Every one who has observed the equality of the seating in Corinthian Hall . . . knows how faithfully he avoided . . . discrimination. The mechanic who came in his shirt-sleeves and with unshaven face, was served as well as the man whose bosom flashed with diamonds."[53] The elite eulogists enjoyed this "democratic" premise without noting the social dislocations that had also accompanied the reclamation project. It was in the spirit of democratic community that the eulogist praised William Reynolds as a public figure: "representatives were gathered here from every class of society." Then, without irony, he listed a bourgeois-heavy interpretation of "society": "merchant, lawyer, banker, professor, teacher, millionaire, and day laborer."[54] The idea of the merchant as the carrier of culture and morality as well as economic productivity was delivered with great flourish befitting funereal respects, and yet its rhetoric meshed with the early nineteenth-century rhetoric of city building too neatly to be coincidental. "It seems to me that the noble body of men who up to this time have given commercial credit, moral tone and an honorable reputation to our city, is fast passing away," worried one writer.[55] A local newspaper noted that the death "will justly be regarded as a calamity to our city," as Reynolds had been "intimately associated not only with the business interests of Rochester, but no less closely with its moral and educational institutions."[56]

Just as merchants had privatized public space, now the private person of the merchant had become a public figure. "His death is a public loss," mourned one newspaper; William Reynolds was "eminently a public man," wrote another.[57] Testifying to the public service of the man, Reverend Bartlett noted that, "were it our custom to hang our dwellings with the drapery of mourning on occasions when a noble, pure-minded, large-hearted, honored citizen had passed from among us, I doubt whether a single home would be without the symbols of bereavement."[58]

For the funeral the Arcade and Athenaeum were added metaphorically to the official list of mourners, draped inside and out in black crepe to "illustrate his tenants' profound love and attachment for their friend and beloved, indulgent landlord."[59] The black crepe of mourning was only fitting because the news of his death "shrouded the hearts of all his tenants in the drapery of sorrow too painful to be described."[60] Flags were hung at half-mast, and the tenants placed a portrait of Reynolds on the rear wall of the bedecked Arcade as if it were a shrine. An "arch of gas jets" illuminated the portrait by night, and an "arch of evergreens and immortelles" replaced it by day.[61] In a rather remarkable turn of events, businesses closed for the funeral of a merchant who had made business his business.

The Reynolds Arcade and Athenaeum showed the depth of the merchant's ability to create a genteel tableau of commercially conditioned space that sorted people as well as activities. We see in the Arcade buildings both the fullest realization of a complex, commercial district within the sorted city and, at the same time, an entity that transcended a merely commercial identity by incorporating functional and rhetorical claims to public space. At the same time, we can see how the patrons of the buildings became lionized themselves as the benefactors of these public monuments and public good for the entire city. At the end of the Arcade hall were three sculpture niches containing busts of Abelard Reynolds and his two sons, who thus joined the pantheon of cultural heroes displayed in the reading room sculpture gallery.

SEVEN

Transportation and the Changing Streetscape

Although both the ideal and the practice of sorting remained strong, the sorted city itself was never a static entity. Change was inevitable, indeed desirable, as new and nimble mercantile cities jockeyed for prominence within their expanding economic orbits. The process of sorting was in constant flux with overlapping, rather than hard-edged, stages of development. As the cities matured, changes in transportation systems powerfully influenced the precise form of the sorted city. In some cases transportation developments unraveled a sorted district; in others it produced a finer grain of sorting.

Transportation improvements presented a paradoxical set of conditions for a maturing city. The economies and layouts of the new cities clung to their roads and canals, yet new modes of transportation did not always accord with emerging ideals of genteel public space. The architecture and clientele of milling, warehousing, and shipping had certainly abetted the merchants' pursuit of economic productivity, but these same improvements did not always mesh with the merchants' bourgeois impulse to order and beautify the cityscape. The enlargement of the Erie Canal in the 1830s and the insertion of railroads from the 1830s to 1850s into the new cities presented new opportunities and challenges. With each incipient transportation improvement residents could anticipate an expansion of the "working" landscape within a district increasingly characterized by genteel retailing or white-collar professional offices. In sum, economic improvement collided with social improvement.

Canalscapes

The Erie Canal, which had been a source of wealth to the local economy, ended up victimizing many of its champions. At first the canal was seen as an unmitigated bonus to its urban hosts. True, the canal altered the original configuration of Rochester's sorted city—it cut through the original southeast milling quadrant in 1821, sealing a trend already afoot to concentrate milling northward to the annexed Middle Falls. But this initial dislocation benefited the city's larger economy at little individual cost. In 1828 the Rochester newspaper publisher Everard Peck exulted that "the artificial river is thronged with vessels bearing to market the products of our highly favored country & returning to use the luxuries and comforts of other climes."[1] By 1836 the canal richly filled the coffers of both the state and Rochester, prompting great local pride that Rochester's earnings were "larger than that of any place west of the Hudson" and making the city's "controlling influence in the navigation of the canal" eminently logical to local eyes.[2] The statewide success of the Erie Canal was measurable in the continual increase of traffic along the forty-foot-wide, four-foot-deep canal. In 1835, with the support of many New York merchants, the state legislature voted to increase the volume of trade by enlarging the canal into a seventy-foot-wide and seven-foot-deep channel.[3] The following year a canal committee urged that the project begin: "In consideration of the strenuous efforts now constantly made to divert trade and travel and between east and west through canals and railroads in other quarters rival to those of this state, we feel it to be due alike to the welfare of this state and to our own interest to aid in arousing general attention a subject of such vital consequence as the ENLARGEMENT, *with all practicable speed,* of our GREAT NAVIGABLE HIGHWAY."[4] Local merchants were feeling the pressure from rival trade centers served by the incoming railroad and were beginning to suffer from what would prove to be a serious financial panic in 1836 and 1837. Many prominent Rochester businessmen, including merchants, millers, and mayors, championed the proposed expansion to protect the city's economy.

The canal expansion, including a wider and sturdier aqueduct of Onondaga limestone, was initially a cause for celebration. At the 21 April 1842 opening of the enlarged canal, a small local ceremony was arranged. Booms blocked both ends of the aqueduct, forcing a ceremonial pause in traffic before formal entry into the city. At the far side of the river, locals jockeyed their canal boats for a position at the end of the official honor parade. The Child's Basin merchant miller Ebenezer Beach, however, was not to be denied his place in the sun. Before the canons had finished their reverberations, he had whipped his team of tow horses into action, raced to the front of the line, snapped the gate booms aside, and was the first to cross.[5]

The rush to the canal, however, proved less of a boon than had been hoped. Because it had been financed, built, and administered by the state, the local government actually had little control over canal affairs, including the issue of "takings." Canal expansion consumed valuable property right in the middle of the city. As part of the mitigation process, the state hired surveyors to assess the damages to lot holders along the route and paid damages where it deemed appropriate, particularly in the case of land takings. The canal board often ruled, however, that damages from lost footage were more than compensated for by the increased value of being on the improved canal and argued that the shaved lots "are valuable in proportion to the extent of the business front and do not diminish in value in the same proportion as the depth is diminished."[6] Local property owners were not persuaded. A row of businesses just beyond the Four Corners on Sophia Street claimed excessive injuries from the loss of property taken up by the new canal embankments. In 1841 the tavern keeper Jonathan Swift angrily requested additional reparations from the New York canal board: "it is a palpable piece of injustice to take away a man's property [and] then screw him down . . . by officers sent from a distance who can have no actual knowledge in regard to its value at the place."[7] The merchant miller Hervey Ely, another Sophia Street property holder, was similarly disgusted by the process. Being told by the state canal board that his shaved property would benefit from the enlargement of the canal, he retorted that he already benefited from the existing canal and did not need the enlargement in the first place.[8] The board did recognize that double-fronted buildings, "with two business fronts one upon the canal and the other upon the street," were particularly vulnerable, however, since they required depth to preserve the integrity of the bifurcation.[9]

Not only did the widened canal consume more land in the city, but also its abutments changed the whole landscape. The new cast-iron superstructure, or "raiseway," which flanked the 1842 aqueduct and canal cut the city in half more thoroughly than the first canal had. Originally, low footbridges connected either side of the city, but now the new structure was less easily pierced by connectors. A frustrated resident decried the "high iron columns, cross-pieces, caps, couplings, bars, bolts, nuts, ties, beams, braces, buttresses, abutments and other paraphernalia" that cut Exchange Street in half, separating north and south.[10] Many businesses were already in trouble, having never fully recuperated from the 1830s financial crisis, but those on the south side of the raiseway found themselves cut off not only from the Four Corners but also from the emerging secondary center at the rail station further north, and they never recovered.

Railroad improvements definitely added insult to injury. In 1837 the first of three competing railroad lines entered Rochester. By 1853 the New York Central Railroad (NYCRR) consolidated the three companies in the city and opened a showy, iron-roofed station.[11] With the passenger station near the main falls and the freight depot at the edge of the Frankfort Tract's central square—the same square that had lost out on its bid for the Monroe County courthouse in 1822—the railroad boosted the commercial and industrial potential of the area (Fig. 3). In addition to prompting a high-style hotel, the new passenger station led manufacturers and wholesalers to build new storehouses and mills nearby, effectively creating a new transport node that surplanted the old canal-based one. New and improved buildings reclaimed what had been a low-rent district and, in the process, created the kind of cityscape in which city builders took pride.

Child's Basin was in eclipse. Establishments along the basin and Exchange Street failed, and owners began a clear pattern of disinvestment and abandonment. As businesses closed, property values plummeted. One bankrupt Exchange Street merchant offered his New York creditors his property as security, but they refused to take it at any price; it was a building, they said, which "could not be rented out for anything."[12] The fortunes of the Rochester House hotel paralleled those of Exchange Street. The Rochester House had enjoyed exceptional prestige as well as patronage. Here Augustus Strong and other local members of the Orion Club's "galaxy of stars" had met.[13] When the 1842 aqueduct was completed, the Rochester House found itself on the wrong side of the raiseway. Business kept slipping. By 1849 the Rochester House's tenants included an African-American shopkeeper selling secondhand clothes from one of the stands. The building had so degenerated that it was not repaired after a fire in 1853. Between 1856 and 1858 the Rochester Industrial School operated out of the worthless building. Its derelict condition was deemed suitable only for the dregs of society: the children who "from the poverty or vice of their parents, are unable to attend the public schools, and who gather a precarious livelihood by begging or pilfering."[14] Eventually, the abandoned hotel became a home for squatters. Edwin Scrantom eulogized the building where the Orion Club had once admired its own wit:

> The martins and swallows took up their summer abode in the cornices of the portico outside, and inside it was given up not only to the owls and bats, but to humans more wakeful and wicked, and its richly furnished drawing rooms and chambers, from the bridal chamber to the inviting family rooms, were changed to the bare floors and squalid conditions of the shiftless and

abandoned, till they descended to "holes," where the occupant stayed, but with all the comforts ending with pig-pens and dog-pens ... a perfect realization of degradation and poverty, that had its representative from almost all grades of human society who had found their way through all the phases of indulgence and excess.[15]

Scrantom regretfully recalled that, "when the Rochester House was in its glory, Exchange Street was *the street* of Rochester."[16] But those days were past, and there was little economic incentive or social will to make improvements. Child's Basin itself closed. Jonathan Child, however, had hedged his bets. As owner of the Pilot boat lines, he had supported the troublesome canal enlargement, but he was also vice president of the Tonnewanda Railroad and remained immersed in the bonanza of the transport business.

During this same period Abelard and William Reynolds were striving to maintain the centrality of nearby Buffalo Street as the eminent retail street. Although they faced challenges from State Street and sites across the river, they did not suffer from the widening of the canal or the loss of Child's Basin. Although they continued to fight architecturally and urbanistically to reclaim the Four Corners for their genteel commercial vision, merchants on the cut-off south side of the canal had less cause for optimism. The bustle of milling, freight, and deal making at Child's Basin was fading, and that district, once so carefully protected and deliberately sorted, began to unravel.

The canal expansion did, however, contribute to the urban aesthetics of improvement with the state's construction of new weighlocks. The first weighlocks were simply weigh chambers at canal grade. Boats were weighed by measuring the amount of water they displaced, and tolls were assessed accordingly. The Syracuse weighlock was constructed just a few blocks east of Clinton Square, its conspicuous location reflecting its prominence in reviving the entire city-building venture. The first Rochester weighlock lay just east of Child's Basin but was relocated to the east side of the river in 1852. During the early 1850s the state rebuilt its Syracuse and Rochester weighlocks into showy Greek Revival edifices that freely adapted the classical idiom within utilitarian buildings. As befit a government edifice, both the canal and street facades were monumental and authoritative. The canal side featured a Greek Revival portico on squared Doric piers of Onondaga limestone whose covered arcade sheltered the docked boats (Fig. 26). The street side addressed the public streetscape and featured quality tight brickwork surmounted by a dignified pediment. Already substantial pieces of architecture, these weighlocks' landscaping further participated in the wave of midcentury refinements. In the new spirit of beautifying the cityscape a Syracuse newspaper lauded the new landscaping of the front yard of the weighlock, including "an ornamental iron railing, which will greatly improve the appearance of that justly elegant structure."[17] Transportation and aesthetic improvements renewed the dignity of the working canalscape, but they could not reverse the canals' competitive decline precipitated by the railroad.

Railroadscapes

As with the canal, the railroad helped put a city literally and figuratively on the map. A railroad solidified a city's position as a transportation center in the greater regional and national trade networks, bringing profits to its host city. Just as the Erie Canal had accelerated the settlement of Syracuse, the railroad heralded its next leap. In 1839, when the railroad clattered into downtown, a national magazine used it as an excuse to highlight the Syracuse as an "enterprising, enlightened village . . . destined to become a great inland city."[18] In 1853 the New York Central Railroad consolidated seven rail companies serving segments between Albany and Buffalo into one new line. By adding new trackage and eliminating others, the NYCRR redirected the original railroads' purpose from linking New York markets to sharing a trans-sectional line connecting New York City to Chicago. Syracuse and Rochester remained on the main trunk line and prospered as part of the expanded, national, mercantile chain; in contrast, Auburn, Cayuga, Geneva, Canandaigua, and Attica were relegated to branch lines.[19] In 1856 a puffing train crossing Salina Street was the backdrop for another national magazine's boosterish account of the commercial progress of Syracuse (Fig. 52).[20] Municipal authorities welcomed the railroad into the center of the city because its economic advantages of convenient and affordable transportation seemed to outweigh the disadvantages of noise and dangerous grade crossings.

In the new economy of railroad and commerce, the antiquated vestiges of the old economy, including canal basins, millponds, and minimally productive salt vats, were reenvisioned as sensory irritations that also impeded commercial real estate development.[21] Once symbols of opportunity, they were now susceptible to charges of being a public nuisance. One of the economically stagnant turning basins east of Clinton Square was deemed a nuisance and filled in. It was replaced by a new symbol of the civic weal—a combination town hall and market hall that continued the practice of double-fronted buildings by backing onto the canal and fronting Water Street.²²

Inserting the Railroad

The 1830s railroad stations adapted the Greek Revival portico into a large roof covering the tracks, trains, and passengers. The 1839 Syracuse station built by Daniel Elliot, who had also constructed new business rows in Hanover Square, was a wooden structure forty-nine feet wide and twenty feet high, with double doors at either gable end through which ran a double set of the train tracks. Passenger platforms flanked the tracks, and raised galleries held a mezzanine-level waiting room, ticket office, superintendent's office, and a telegraph office. A cupola perched on the ridge provided ventilation and signaled the public status of the building below, much like the architecture of the combination market and town hall that would be built in the next decade.²³

During the 1840s the Syracuse station was reconfigured into a compact-and-sorted space that was aesthetically and socially refined. The Greek Revival station was itself segmented. The new ticket office fronted Salina Street, with rooms behind it holding a hall and a gentlemen's sitting room that faced the action on the tracks. Behind these rooms was the ladies' sitting room, the same room where Augusta Rann sat in modest withdrawal. The seclusion from the bustling tracks was not entirely sweet because the room faced the backyard privies. Yet even this close positioning may have provided the privacy that ladies needed in order slip away to the public conveniences.²⁴ Food was the great bridge between the sexes; both sitting rooms connected to the eating room. Located behind the railroad station, the Citizen's Coffee House and Ice Cream Saloon followed the gendering of public space which was occurring in the train station and major hotels. In 1843 the owner of the Citizen's Coffee House and Ice Cream Saloon forged gendered spaces within his commercial establishment. The "refreshment room" on the ground floor was fitted up "after the latest in city fashion" for dining. The second story was divided into a more private suite of public rooms—a "Private Saloon for Ladies" outfitted with the requisite piano and another mixed-gender room where "gentlemen accompanying Ladies" could be served.²⁵

The location of the train station had a major effect on the redevelopment of the commercial district. The effects of the decisions about where to run the tracks and construct the station rippled through the sorted city. The route of the railroad and the location of the station carried the potential to open up new areas for trade and thus to alter the status quo of the original commercial district's centrality. On the one hand, the railroad served the same function as the canal, and thus it could have been inserted into the urban fabric in virtually the same place. Certainly, there were proponents who argued this basic point. In Syracuse many businessmen advocated bringing the railroad in along Water Street, thus linking it to the hub of the Erie Canal dock and the Clinton and Hanover Square trade center.[26] But real estate speculators turned events in their favor. Vivus Smith and John Wilkinson, the railroad commissioner and president of the Syracuse and Utica Railroad, respectively, had other plans. Wilkinson had been playing the Syracuse real estate market for decades; he was the one who had opened his law office on the corner of Washington and Salina streets back in 1819, when the southerly site was considered out of town. Wilkinson, it turns out, was laying the groundwork for his own real estate speculation. Twenty years later he routed his railroad company through his holdings. The tracks ran down Washington Street, and the new station was across the street from Wilkinson's office, rebuilt as the prominent Globe Block.[27]

With the train stopping at the corner of East Washington and South Salina streets, South Salina Street became a magnet for more businesses. The 1839 railroad spawned a new generation of building and activity within the vicinity. Without the introduction of the railroad in Syracuse, the expansion of the commercial district could more easily have headed northward. Indeed, the salt production center of Salina lay to the north, as did the compromise Onondaga County courthouse, until it was relocated to Clinton Square in 1856. After the routing of the railroad, however, the bulk of new investment turned southward down the spine of South Salina Street.

Encouraged by the railroad, merchants, hoteliers, restaurateurs, and owners of large building blocks staked out South Salina Street. The congregation of business blocks along South Salina Street marked out the expansion of the commercial district below the train station. Hotels with dining facilities sprang up specifically to serve the time-pressed railroad passenger. Merchants tapped into this new flow of traffic and invested in upgraded commercial facilities. In doing so, they expanded the boundaries of the older commercial district. It was in this real estate climate of the 1850s that the Italianate Pike Block replaced a prosaic farmers' inn at South Salina and Fayette streets and Henry Dillaye launched his ambitious Washington Block project across the street (Figs. 28–29).

Although the railroad station was obviously used as part of the transportation system, the building played a much larger role, serving commercial and public purposes as well. Commerce and the railroad station were tightly linked in the larger scheme of communications. Just as Reynolds Arcade functioned as communications central for Rochester, the Syracuse rail station assumed similar functions, with the telegraph office and the post office clustered around the station. Not coincidentally, the postmaster at the time was John Wilkinson.[28] As part of the public landscape, the station drew activity away from Clinton Square and its institutions. The station's downtown location made it part of the daily landscape as well as a convenient hall for "whistle-stop" speeches. The station supplanted the balconies of the Syracuse House hotel as the public forum, and traveling dignitaries delivered their orations from the galleries and platforms. A public ceremony reiterated the close relationship between the railroad, communications, and civic pride. In 1858 the Syracuse station was the site of a celebration of the country's laying of the Atlantic cable. At the given signal, thirteen steam engines gathered at the station blew their whistles in salute. The effect was not what the planners had imagined: "The effect upon the listeners was most appalling, producing an electric shock that made the strongest turn pale; some women and children were thrown into convulsions and did not fully recover from the shock for days."[29] The shriek of the whistles converted that celebration of technology into a mass trauma, but it also sealed the connection between the station as an urban institution and its link to long-distance communications.

The railroad helped Syracuse develop a unique market niche during the third quarter of the century. Known as the "Central City" for its central location in the state and crisscrossed by several railroad lines as well as the canal, Syracuse was easy to get to. Entrepreneurs capitalized on the city's centrality and the rising practice of state and national organizations holding conferences by building large halls that would accommodate conventioneers. As a Rochester resident ruefully reported, "In the matter of public halls, Syracuse is well provided, having three halls of the seating capacity of our own Corinthian Hall."[30] An eastern New Yorker who attended a church conference in Syracuse noted that "Syracuse is a city of conventions and has provided ample halls for their accommodation. We have seldom enjoyed a session more than the one just closed."[31] He concluded that "the Central City is a good place to go to—to be used well."[32] Retailed stationery letterheads featuring a vignette of the train station further promoted the city and its convenient access.

The railroad confirmed the district's commercial focus, but trains also added to its congestion, with all their clutter, smells, and sounds. A traveler passing through Syracuse in 1857 commented on the disjuncture between the merits and drawbacks of a robust, noisome, and dynamic rail landscape: "Upon first emerging from the cars it is not strange that the tourist's impression is often disagreeable, but the activity everywhere is perceptible in and around the spacious and well arranged station, the crowded stores; and the busy expression which each passing countenance wears, will immediately put to flight any preconceived notions you may have formed on the score of the insignificance of the Salt City."[33] Although the visitor ultimately made peace with the lack of gentility in the downtown, he nonetheless felt compelled to mention it. The railroad was an economic asset, but merchants and the residents began to question how well a railroad fulfilled that role in a downtown location.[34] Downtown trains clogged the streets, impeding the district's efficiency and convenience. As part of the negotiations to usurp a public street for a privately owned railroad, the city required the Syracuse and Utica Railroad Company to construct a sewer line along Washington Street and to plant trees on both sides of the street. But, despite the opportunities that service industries saw in snagging travelers, not all business prospects were well served by the trains.

Relocating the Railroad Station

The movement to relocate the train terminal was part of a larger convergence of commercial motives, bourgeois sensibilities, and civic-mindedness precipitating public reassessment of the urban landscape. Three locations in the city had become a focus of civic attention: the commercial extension of South Salina Street, the antiquated millpond and saltworks to the west of South Salina Street, and the hillock between Syracuse and Salina near the old compromise courthouse on North Salina Street. A new wholesaling pocket emerged at the western edge of the commercial district as a response to all three foci.

As part of the maturation of the economy, merchants were increasingly able to develop separate retail and wholesale enterprises, an economic practice with spatial repercussions. Retailers still needed good access to the transportation lines that carried in their stock, but they did not need immediate proximity to heavy-transport lines in the way wholesalers did. In contrast, wholesalers required convenient access to transportation but were less rooted to the pedestrian-oriented commercial center. Enjoying greater capitalization, greater transportation options, and greater product specialization, the merchants were beginning to conceive of two very separate districts for the two separate branches of their business. The commercial community was willing to sort at a finer grain than before. As long as the railroad stayed put in the center of downtown, so, too, would the wholesale district.[35]

These threads all converged at the old Syracuse Company holdings of the Onondaga millpond and the state's Syracuse Salt Works, which adjoined it. In the new climate of refinement the seasonal activities at the saltworks and millpond became targeted by those wishing to replace the economically negligible industrial pocket with developments more suited to a city center. Arguments of health and market efficiencies were presented. The baneful landscape was not only noxious but also a grossly unproductive use of downtown real estate.

In 1842 municipal authorities first petitioned the state to release its hold on the saltworks west of the Onondaga millpond, noting, "it is a very desirable tract for building purposes," which would bring large revenues to the state if auctioned for sale.[36] A state surveyor agreed, noting that if the works were eliminated then Fayette Street, "now one of the most beautiful streets in the city," would be connected with its mate across the saltworks, forming a "continuous street of nearly four miles," which was now interrupted by a saltworks oddly located less than five minutes' walk from the elegant Syracuse House hotel.[37] In 1848 a group of citizens petitioned the Syracuse Company to fill in its Onondaga millpond property. The principal, John Townsend, refused, defending the company's retention of the pond, conceding it would implement measures to negate the nuisance, and reminding local leaders that the Syracuse Company had a long-standing tradition of active urban investments in the city. "I am very sure no person has taken a greater pride and done more than I have to promote the prosperity and general improvement of the city of Syracuse," he argued.[38] Syracuse, however, had just been legally chartered as a full city, and one if its first acts as an independent entity was to flex its municipal authority. E. W. Leavenworth, the one-time president of the village who had championed the creation of residential parks and tree planting, was now elected mayor of the city. The city council invoked its authority to regulate nuisances, and Leavenworth had the offending section mapped and laid out into a elegant oval park known as Jefferson Square.[39]

The city literally had the dirt for fill as a result of grandiose political schemes. Rumor had it that the capitol might be moving away from its colonial site in Albany to a more central location. Syracuse's excellent rail, canal, and turnpike connections raised the hopes of locals that their Central City would be picked. Leaders identified the eminence of Prospect Hill, not far from the compromise courthouse, for the capitol building and graded the site as its foundation or terrace.[40] They carted the dirt away from the hill and dumped it into the millpond. In one earth-moving gesture the city had created a plateau for a capitol and removed the fetid nuisance of the stagnant pond.

The reclamation of the millpond opened up a new zone for redevelopment and commercial specialization which went hand in hand with the railroad. In 1853 the Binghamton railroad opened a station at the newly developed Jefferson Square, and one resident crowed, "it has had quite an effect on vacant property in that vicinity."[41] In 1857 the state commissioned Horatio Nelson White to design the Fifty-first Regimental Armory for the center of the oval park, adding another architectural landmark to the city and one that needed close access to shipping lines.[42] In 1868 the New York Central Railroad added a spur from Washington Street to a new terminal next to the armory and tore down its old station at Washington and Salina streets. The trains still ran through Washington Street, but they no longer stopped at the Salina Street corner. Wholesalers, such as McCarthy's Dry Goods wholesaling operations, congregated to the new railroad site around Clinton and Fayette streets, where they built large, showy warehouses. A Rochesterian passing through Syracuse noted that the railroads were "a great source of wealth, and bring to it trade which enables some of its wholesale business houses to report incomes in the fourth and fifth cyphers."[43] By the third quarter of the nineteenth century warehouses and wholesalers ringed the Armory in Jackson Square, and all tapped into the easy access of the expanding rail lines, forming a new wholesaling district within the expanded commercial district (Fig. 56).[44]

Back on Salina Street, retailers filled in the vacuum of the old station site. The removal of the train station and associated freight office in 1868 opened up new spaces for new, pretentiously genteel business blocks, many of which are still extant, including the 1869 Granger Block and Larned Building and the 1876 White Memorial Building, a fashionable High Victorian Gothic, mixed-use commercial structure. The construction of high-style commercial buildings with their retail and professional offices articulated the heightened refinement of the commercial district. The introduction of the railroad had first helped to stimulate the retail and wholesale aspect of the commercial district; its removal provided more stimulus but this time to forge separate retail and wholesale subdistricts.[45]

The McCarthys on the Move

The geographic strategies of the Salina-based McCarthy family's mercantile business illustrate the appeal, domination, and subdivision of Syracuse and its commercial district as well as the merchant's role in shaping the sorted cityscape. The village of Salina, older than Syracuse and ensconced in the salt production business, was now struggling to keep up with the canal-hugging upstart town to its south. The county commissioners had earlier compromised between the two rival settlements with the Onondaga County courthouse compromise, but the next-generation merchants were not so impartial. In new city terms the McCarthys were an old Salina family. Thomas McCarthy had emigrated from Ireland to Salina around 1808, prospered in salt making, and rose to the positions of merchant, bank president, village trustee, and church leader. Father and son ran a successful mercantile establishment on Park and Free streets between 1834 and 1846, when the father retired from business. That year the son, Dennis, disengaged from his father, his village loyalties, and the rest of his hometown merchants. Despite his father's legacy and his family's powerful associations with Salina, Dennis McCarthy weighed the balance of trade and calculated Syracuse the winner. Despite the senior McCarthy's displeasure, Dennis closed the Salina shop and reopened the dry goods business in Syracuse.

McCarthy's new business locations followed the course of dry goods row. First, in 1846 Dennis McCarthy moved from Salina to a storefront in the Empire Block on the north side of Clinton Square in dry goods row (Figs. 5, 24). The allure and convenience of a canal location, however, was losing its hold in light of the expanding commercial district moving southward to the new railroad station. In 1850 McCarthy moved away from North Salina Street to South Salina Street, renting the four-story Dillaye business block at the corner of East Fayette Street (Figs. 5, 54).

McCarthy was not finished. By 1872 he had built his own large store, which housed both wholesale and retail departments carrying a huge assortment of dry goods. In 1876 he hired Archimedes Russell to design a separate wholesale emporium near the emerging wholesale district on the reclaimed millpond of Jefferson Square (Figs. 5, 56). Due to his shrewd business tactics and commercial success, McCarthy came to be known as the "A. T. Stewart of Syracuse," a reference to the famous New York City department store magnate whose success had also inspired the Reynolds family to upgrade their Arcade.

McCarthy set the trend. Shortly after he left Salina, the Lynch Brothers dry goods business followed. The dry goods merchants Williams & Hutchinson similarly moved from the Empire Block, first to the Franklin Row on Hanover Square and then to the new Washington Block on South Salina. Moving merchants were commonplace in nineteenth-century cities, so much so that the business publisher Freeman Hunt felt compelled to advise shopkeepers to stick with their locations. "A rolling stone gathers no moss," Freeman repeated, and "three removes are as bad as a fire."[46] Hunt's advice was soundly rejected by many Salina merchants, who judged Syracuse the rising city. In 1848 Salina and Syracuse had joined together as one incorporated city, erasing Salina as an independent political entity and confirming the rise of Syracuse. In 1853 the residents voted the Salina expatriate Dennis McCarthy to be the mayor of Syracuse, symbolizing the closure of Salina's independent political and commercial status. The ultimate resolution of these defections can be seen in the state of the Salina left behind (Fig. 53). Salina's snaggletoothed Wolf Street revealed a failed city trailing behind its Syracuse competition (Fig. 54).[47]

Reviewing the Sorted Cityscape

Two midcentury bird's-eye views, an 1852 view of Syracuse made by Lewis Bradley and an 1853 view of Rochester made by John W. Hill, interpreted and broadcast the evolution of the sorted mercantile city (Figs. 55, 57).[48] Both lithographs are high-quality, detailed views of what had become moderately large, prosperous cities. By 1858 Syracuse had a population of over 25,000 residents, and the city's real estate was valued at $6,381,356 and personal property at $1,765,463. Rochester's population was nearly 44,000, and its real estate was valued at $9,362,408 and personal property at $2,582,565.[49] The views themselves are sizable, each measuring approximately two feet by three feet, both depicting impressive cities that sprawled past the frame of the view. The production of such urban views paralleled the rise in the number of and, more significantly, the interest in the phenomenon of cities. Lithographs of American cities proliferated during the second half of the nineteenth century, when at least twenty-four hundred villages, towns, and cities were the subject of enterprising view makers.[50] Once relegated to the background as a backdrop for the main subject of the painting or print, the city now became the subject itself.[51]

The production of urban views gave the middle class one more opportunity to act as urban boosters who shaped and promoted the mercantile city and its representations. For all the curiosity about distant cities and pride in one's own town, distant lithographers would not invest in the time-consuming project without private subscriptions from within the targeted city. Just as elite merchants rose to the call to be architectural patrons of the city, they similarly assumed the mantle of urban promoter by placing bulk orders for city views. But, unlike the case with expensive architectural commissions, it was possible for the more common resident to become a patron of the city by ordering a single print. On occasion a subscriber would order a handful of views or local retailers would order stock—Dewey's News Room in Rochester carried Edwin Whitefield's 1847 view of Rochester—but most lithograph subscriptions were promised one at a time.

Bird's-eye views, like any other commissioned portrait, were commissioned by people who wanted a credible but satisfying picture of the esteemed subject. In this case the subscribers to the lithograph project were joined together as an urban family seeking an object for their urban veneration, and the resulting views should be seen as documents that met with their approval. Few portrait sitters commissioned a portrait "warts and all"; the same was true for the patrons of city views. True, to be a good portrait, it had to be sufficiently accurate to be persuasive and recognizable, and each view exhibited meticulous and realistic architectural details.[52] And yet, in this genre, the sun is always shining, the buildings are always in good repair, and the streets are always broad and bustling. The cultural context of their production influenced the existence, content, and perspective of the views.[53] The city views are therefore rich sources of information about bourgeois notions of what made a city worth celebrating.

As mechanically reproduced images, the bird's-eye views could be easily disseminated. After an 1847 view of Rochester was published, the local paper extolled it as "an ornament peculiarly worthy of a place in the office and houses of Rochester men."[54] Similarly, the advance publicity for an 1859 lithograph of Syracuse pushed the view as "suitable for parlor ornaments, hotels, offices and other places."[55] City portraits were not relegated to stay in the city. The centripetal "hometown" focus of the view assumed a centrifugal aspect once the images were flooded into the public realm, typically at the hands of the merchant class and later in the century by railroad interests who posted them in hotels and train stations. The views were often reduced from the large folio size to octavo size, approximately 4.5 by 7.5 inches, for easier handling. The New York publisher Charles Magnus minutely shrank Hill's 1853 Rochester lithograph down to stationery letterhead only a few inches in size. With the stationery letterheads especially, the urban portraits became an easy way of depicting and interpreting the city for a diverse and scattered audience. It is little wonder, then, that the bird's-eye view and the city builder's view converged on the same page. The bird's-eye view was one more tool that the city builders could use to promote their version of the ideal city.[56]

So, what was the image of the mercantile city? Both the medium and the content of the lithographs show the persistence and realization of the mercantile vision that had stimulated the city-building endeavors at the turn of the century. These portraits were not simply pretty and proud pictures of the cities; they were statements confirming the economic and cultural improvements that urbanization had wrought. Far from proving Jefferson's fear that cities were "pestilential to the morals of men," the new cities were held up as agents of positive economic and social progress. The ways in which the Syracuse and Rochester views are framed reflected a particularly booksterish attitude of putting the city, or more specifically the city's businessmen, in the best light. The Syracuse view is part of a whole genre of urban views that is strongly linked to the tradition of landscape painting, with the composition of three horizontal bands: a pastoral image in the foreground, the urban vista in the mid-ground, and a band of nature and sky filling the background. In a sense this lithograph is more like a landscape with a city in it than it is a pure urban view. Lewis Bradley's landscape approach was not a mindless assumption of landscape painting traditions, but the deliberate fusion of urban imagery with nature imagery which could be found in other urban views of the time. The pastoral city was a method of reconciling two contradictory attitudes about the direction of American growth: the yeoman republic and a long tradition of antiurbanism versus the rise of cities dedicated to commerce and manufacturing. In this view Syracuse sits comfortably in its pastoral frame, existing both within and separate from a cultivated landscape. Any negative implications of urbanization were diminished by showing this artificial and spectacularly growing city as coexisting within nature. The resulting urban vision was a hybrid, Leo Marx's "middle landscape," a contrived image that expressed the "national preference for having it both ways ... [and enabling] the nation to continue defining its purpose as the pursuit of rural happiness while devoting itself to productivity, wealth, and power."[57]

Economic productivity in processing the raw resources of the region and in developing an urban commercial economy was obvious. In the full Rochester view Buffalo Street picks up with commercial busyness in the mid-ground, the Erie Canal regularly sports canal boats and warehouses, and in the distance the spray of the Middle Falls is nearly obscured by the dense packing of mills. Even the pastoral Syracuse panorama shows billows of boiling saline waters and rows of solar evaporation tanks ringing Onondaga Lake. Such industrial activities, of course, could not have been profitable on any large scale without transportation improvements, and indeed these requisites are also highlighted—more so in the Rochester view, but the Syracuse view also features two parallel lines of the canal and railroad slicing into the city from the right. Good transportation options combined with natural resources made the commercial aspect possible, and both of these cities appear quite secure within the mercantile network of trade.

Economic improvement was only part of the goal of colonizing the eastern frontier. These cities were also supposed to be conveyors of civilization, and a particularly merchantlike, middle-class interpretation of civilization at that. The Syracuse view visually answers the question that a Syracuse preacher had righteously posed to his congregation in 1846: "Who rules in Syracuse, God or Mammon?"[58] Both, was the answer, and they are interdependent. Indeed, the center of the view features two overscaled, competing building types: the business block and the church. Commercialization supports and suggestively elevates religion, which in turn elevates the tenor of society. Edwin Scrantom mused about Rochester's commingled economic and moral improvement: "The place grows rapidly about us— and instead of the wilderness that frowned here in 1813, a *city* now is here teeming with people and with business, and spires are seen, as it were pointing the wayward and thoughtless to a better & more permanent city, 'it made with hands eternal in the heavens.'"[59] Religion and commerce could be commingled just as logically as industry and commerce could. "It is evident," the business publisher Freeman Hunt rhapsodized, "how much we owe to Commerce, and how greatly we depend upon our merchants, for our means both of social progress and religious effort."[60]

In these rosy views the cities have disciplined slothfulness as well. People appear busy, and even the leisured picnickers on the hills outside Syracuse are quite respectable citizens, properly enjoying the view of urban, agricultural, and manufacturing improvements before them. Tidy rows of houses further suggest a community of prosperous freeholds. The merchant's dread of an underutilized landscape and the personal depravity of the Native Americans or isolated farmers had been replaced by the bourgeois notions of good urban order.

The well-ordered city was a sorted city. The central city, whether defined by the slice of Buffalo Street in Rochester or the tight packing of blocks in

Syracuse, was nonetheless clearly a commercial center. Separate nodes held the processing and refining activities, and separate lines channeled the transportation landscape. The civic landscape was boldly ringed by churches in Syracuse and could be picked out in Rochester through careful examination of the churches and courthouse lining Buffalo Street as it approached the Four Corners. The compact form of the city was as much in evidence as was its sortedness.

The new cities gave vivid testimony to the urbanizing culture of improvement on the eastern frontier. As early as 1829, the journalist William Leete Stone had captured the merchants' pride in the speedily wrought sorted cities. "And this is Rochester! the far-famed city of the West, which has sprung up like Jonah's gourd! Rochester, with its two thousand houses, its elegant ranges of stores, its numerous churches and public buildings, its boats and bridges, quays, wharves, mills, manufactories, arcades, museums, everything—all standing where stood a frowning forest 1812! Surely the march of improvement can never outstrip this herculean feat."[61]

Today historians may be surprised by the presence of a sorted cityscape in the compact, new, nineteenth-century city. But the examples of Rochester and Syracuse indicate that sorting was a deliberate process that mercantilist and bourgeois city builders undertook from the beginning. Sorting transformed a new settlement into an efficient, well-ordered, and appealing landscape that was built for profit first and bourgeois comfort second. The upstart cities surprised nineteenth-century observers, too, who devised their own creative explanations for the mature cityscapes they saw. Time traveled faster in the hinterland, explained one traveler, and "ten years in America are like a century in Spain."[62] Contemporaries puzzled over the pace of growth of these upstart cities, whose accelerated progress threw off all notions of urban history and timing. They did not, however, question the sorted cityscape that was appearing before their curious eyes. The speed but not the concept of sorting was the marvel of the new city. The nonchalance of the nineteenth-century public to this socio-spatial pattern suggests that the merchant's new and sorted cityscape was a far more ordinary urban type than historians have considered.

Notes

ONE
Vernacular Urbanism and the Mercantile Network of New Cities

1. Alan Taylor, *William Cooper's Town: Power and Persuasion on the Frontier of the Early American Republic* (New York: Knopf, 1995), 4.
2. John W. Barber and Henry Howe, *Historical Collections of the State of New York; Containing a General Collection of the Most Interesting Facts, Traditions, Biographical Sketches, Anecdotes, Etc.* (New York: S. Tuttle, 1841), 395.
3. Ibid. The phrase meant an utterly lifeless environment that could not even support keen birds of prey. See, e.g., Jacques Milbert, *Picturesque Itinerary of the Hudson River and Peripheral Parts of North America* (1828–29) (Ridgewood, N.J.: Gregg Press, 1968), 130, in which he reported that "wherever vegetation growth attains special vigor, living creatures are weak," and so only birds of prey, particularly owls, were able to pierce the dark canopy of overgrowth in order to hunt. A landscape that would make an owl weep was a deeply inhospitable one. Contemporaries similarly commented that the "swamp foliage was so thick, and darkened the atmosphere to such an extent, that the owls, mistaking day of night, could be heard hooting." Thurlow Weed, "Stage-Coach Traveling Forty-Six Years Ago," *Galaxy* 9:5 (May 1870): 602.
4. William Leete Stone, "From New York to Niagara, Journal of a Tour . . . in the Year 1829," *Buffalo Historical Society Publications* 14 (1910): 221.
5. My use of the term *urban threshold* is indebted to Stuart M. Blumin, *The Urban Threshold: Growth and Change in a Nineteenth-Century American Community* (Chicago: University of Chicago Press, 1976).
6. Sam Bass Warner Jr., *The Private City: Philadelphia in Three Periods of Its Growth* (Philadelphia: University of Pennsylvania Press, 1968); *Streetcar Suburbs: The Process of Growth in Boston, 1870–1900* (Cambridge: Harvard University Press, 1978), 11–29, 63, 153; *The Urban Wilderness: A History of the American City* (New York: Harper and Row, 1972), 55–112.
7. Warner, *Urban Wilderness*, 82. Twenty-five years later the historian James Lemon elaborated that in the walking city "people were not quite 'jumbled to-

gether' in chaos, or even rough equality" and that "urban space was structured politically, socially, economically, and hence materially." James T. Lemon, *Liberal Dreams and Nature's Limits: Great Cities of North America since 1600* (Toronto: Oxford University Press, 1996), 78. Spera has also suggested that the walking city of Philadelphia may have been more functionally sorted than previously thought. Elizabeth Gray Kogen Spera, "Building for Business: The Impact of Commerce on the City Plan and Architecture of the City of Philadelphia, 1750–1900" (Ph.D. diss., University of Pennsylvania, 1980). Although New York City had exploded beyond the presumptive size of the walking city by 1850, scholars have explored the segmentation of land uses and the architectural expression of its economic districts at midcentury. Mona Domosh, *Invented Cities: The Creation of Landscape in Nineteenth-Century New York and Boston* (New Haven: Yale University Press, 1996). For a critical appraisal of the lacuna of urban typologies and greater awareness about the tripartite segmentation of warehousing/transportation, commerce, and residential fabric, see Michael P. Conzen, "The Morphology of Nineteenth-Century Cities in the United States," in *Urbanization in the Americas: The Background in Comparative Perspective,* ed. Woodrow Borah, Jorge Hardoy, and Gilbert A. Stelter (Ottawa: History Division, National Museum of Man, 1980), 119–41. See also Diane M. Shaw, "Sorting the City: Urban Landscapes in Upstate New York, 1810–1850" (Ph.D. diss., University of California at Berkeley, 1998).

8. Blackmar has famously re-walked the socially stratified landscape of the walking city. Elizabeth Blackmar, "Rewalking the Walking City: Housing and Property Relations in New York City, 1780–1840," *Radical History Review* 212 (1979): 131–148. By showing social stratifications in housing in New York and Philadelphia, social historians such as Elizabeth Blackmar and Stuart Blumin have amended the generalization that the early-nineteenth-century walking city was residentially undifferentiated, or that social distinctions were spatially irrelevant. Indeed, their work has shown that even within one urban block residents ranked social and physical space. Elizabeth Blackmar, *Manhattan for Rent, 1785–1850* (Ithaca: Cornell University Press, 1989), 72–108. See also Stuart M. Blumin, *Emergence of the Middle Class: Social Experience in the American City, 1760–1900* (New York: Cambridge University Press, 1989), 20–25, 48–51, 151–55. The social historian Richard Bushman expands the class-based interpretation of a sorted landscape by moving beyond the house. Bushman examines the way that gentility became spatialized in the cities. He argues that Americans' tendency to sort out urban space dates back to the eighteenth century, when the urban gentry molded the built environment to suit its specialized needs. In cases in which material changes were impossible, elites "mentally zoned the city into spaces for themselves and others for the vulgar public." Bushman sees separation as part of the gentrification of American society, with the class-demarcated physical landscape buttressing a class-based cultural landscape.

9. E.g., "As cities grew, their business districts became more defined, but here too residences and primitive factories remained interspersed with the stores, banks, and offices." Howard P. Chudacoff and Judith E. Smith, *The Evolution of American Urban Society* (Englewood Cliffs, N.J.: Prentice-Hall, 1975), 76–77.

10. James F. W. Johnston, *Notes on North America: Agricultural, Economical, and Social*, 2 vols. (Edinburgh: William Blackwood and Sons, 1851), 1:204.

11. Frances Trollope, *Domestic Manners of the Americans* [1832], ed. Richard Mullen (Oxford: Oxford University Press, 1984), 333.

12. Anthony F. C. Wallace, *The Death and Rebirth of the Seneca* (New York: Vintage Books, 1969), 210. For the history of the New York western lands, see generally Edward T. Price, *Dividing the Land: Early American Beginnings of Our Private Property Mosaic* (Chicago: University of Chicago Press, 1995), 234–36; William Wyckoff, *The Developer's Frontier: The Making of the Western New York Landscape* (New Haven: Yale University Press, 1988); Barber and Howe, *Historical Collections;* William Herbert Stiles, "A Vision of Wealth: Speculators and Settlers in the Genesee Country of New York, 1788–1800" (Ph.D. diss., University of Massachusetts, 1978).

13. Despite the nineteenth-century fascination with the efflorescence of American cities, urban and architectural historians have paid more attention to the high-order metropolises than to the cities of lower rank. "From an architectural perspective," the most important American cities have been New York, Philadelphia, Boston, Chicago, San Francisco, and Los Angeles, all top ranking in terms of economic, demographic, and cultural regional and national dominance. Richard Longstreth, "Architecture and the City," in *The Divided Metropolis: Social and Spatial Dimensions of Philadelphia, 1800–1975*, ed. William W. Cutler III and Howard Gillette Jr. (Westport, Conn.: Greenwood Press, 1980), 157. There are notable exceptions, and certainly Blake McKelvey's urban biographies of Rochester, New York, stand out for their intensive focus. See, e.g., Blake McKelvey, *Rochester: The Water-Power City, 1812–1854* (Cambridge: Harvard University Press, 1945); and *Rochester: The Flower City 1855–1890* (Cambridge: Harvard University Press, 1949).

14. New York City was the largest, with 123,706. Philadelphia was half the size of New York, with 63,802. Third-ranked Baltimore was not far behind, with 62,738. Boston stood farther back, with 43,298, and New Orleans was a distant fifth, with 27,176. Barbara Shupe, Janet Steins and Jyoti Pandit, *New York State Population, 1790–1980: A Compilation of Federal Census Data* (New York: Neal-Schuman, 1987), 297, 253, 261. "Table 4a—Aggregate Population of Cities," *Compendium of the Eleventh Census (1890)*, 434–35.

15. *Compendium of the Tenth Census (1880)*, xxviii–xxix. By 1850 Syracuse ranked twenty-second of all cities in the United States, with a population of 22,000, and Rochester seventeenth, with 36,000. "Table 4a—Aggregate Population of Cities," *Compendium of the Eleventh Census (1890)*, 434–35; Adna Fer-

rin Weber, *The Growth of Cities in the Nineteenth Century: A Study in Statistics* (New York: Macmillan, 1899), 21, 22. Some of the newly recorded cities were not completely new but were small unrecorded settlements included within larger townships. Oswego, New York, for example, was founded in 1722 as a trading post but did not appear as a city until the 1850 census. Allan R. Pred, *Urban Growth and City-System in the United States, 1840–1860* (Cambridge: Harvard University Press, 1980), 24–26.

16. The three partners shared southern roots and probably first met in Hagerstown, Maryland. Raised in Hillsborough, North Carolina, Nathaniel Rochester (1752–1831) began his varied career as a merchant, speculator in town lots, state assemblyman, manufacturing investor, and overseer of the construction of the local county courthouse, all of which guided his vision of the merchant-led city of commercial, civic, and manufacturing interests. In 1783 he relocated to Hagerstown. William Frisby Fitzhugh (1761–1839) moved from the Maryland family home in Calvert County to Washington County in 1787 and developed a plantation in Chewsville outside of Hagerstown in 1792. Charles Carroll (1767–1823), related to the more famous Charles Carroll of Carolltown, moved to Hagerstown in 1789 from his Maryland ancestral home of Belle Vue. The three men shared other business interests in addition to their land speculation in western New York. In 1807 Nathaniel Rochester organized the Bank of Hagerstown, and the following year both Fitzhugh and Carroll were on its board of directors. By 1810 Nathaniel Rochester had relocated to western New York, near Dansville, and in 1818 made his final move to Rochester. Carroll and Fitzhugh moved to Groveland, near Williamsburgh, New York, in 1815 and 1817, respectively, but neither one ever relocated to Rochester. Robert F. McNamara, "William Firsby Fitzhugh: Co-founder of Rochester," *Rochester Historical Society, Genesee Country Occasional Papers* 16 (1984): 1–36; "Charles Carroll of Belle Vue, Co-Founder of Rochester," *Rochester History* 42:4 (October 1980): 1–28; Blake McKelvey, "Colonel Nathaniel Rochester," *Rochester History* 24:1 (January 1962): 1–23; *Rochester: The Water-Power City*, 24–48; Howard L. Osgood, "Rochester: Its Founders and Its Founding," *Rochester Historical Society Publication Fund Series* 1 (1922): 53–70; William F. Peck, *History of Rochester and Monroe County, New York, from the Earliest Times to the Beginning of 1907* (New York: Pioneer Printing, 1908), 70; Henry O'Reilly, *Settlement in the West: Sketches of Rochester; with Incidental Notices of Western New York, &c.* (Rochester: W. Alling, 1838), 407–16.

17. J. H. French, *Historical and Statistical Gazetteer of New York State* (Syracuse: R. P. Smith, 1860), 488. Joshua Forman lived in Onondaga Hollow, a village of few buildings which provided no urban exemplar, although he attended college in Schenectady and would have known the crooked streets of the nearby capitol of Albany as well. A colleague in the law, John Wilkinson, and Forman's brother, Owen Forman, resurveyed Syracuse in 1819, converting James Geddes 1804 survey into a city plat. W. Freeman Galpin, "The Genesis of Syracuse," *New*

York History 30 (January 1949): 20–26; Joshua V. H. Clark, Onondaga; or Reminiscences of Earlier and Later Times; Being a Series of Historical Sketches Relative to Onondaga; with Notes on the Several Towns in the County and Oswego (Syracuse: Stoddard and Babcock, 1849), 2:85–89; Gurney S. Strong, Early Landmarks of Syracuse (Syracuse: Times Publishing Co., 1894), 311–32.

18. [Nathaniel Hawthorne], "Sketches from Memory," *New-England Magazine* 9 (December 1835): 399.

19. Larry S. Bourne, "Introduction" and "Urban Spatial Structure," in *Internal Structure of the City: Readings on Urban Form, Growth, and Policy*, 2d ed, ed. Larry S. Bourne (New York: Oxford University Press, 1982), 6, 30; Helen Liggett and David C. Perry, "Spatial Practices: An Introduction," in *Spatial Practices: Critical Explorations in Social/Spatial Theory*, ed. Helen Liggett and David C. Perry (Thousand Oaks, Calif.: Sage Publications, 1995), 1–12; Helen Liggett, "City Sights/Sites of Memories and Dreams," in Liggett and Perry, *Spatial Practices*, 243–73; Thomas A. Reiner and Michael A. Hindery. "City Planning: Images of the Ideal and the Existing City," in *Cities of the Mind: Images and Themes of the City in the Social Sciences*, ed. Lloyd Rodwin and Robert M. Hollister (New York: Plenum Press, 1984), 133–47.

20. Setha M. Low, "Spatializing Culture: The Social Production and Social Construction of Public Space in Costa Rica," *Theorizing the City: The New Urban Anthropology Reader* (New Brunswick: Rutgers University Press, 1999), 112.

21. Ibid.

22. Theories about the notion of landscape, which have informed these ideas about cityscape, can be found in John Stilgoe, *Common Landscapes of America* (New Haven: Yale University Press, 1982); and John Brinkerhoff Jackson, *Discovering the Vernacular Landscape* (New Haven: Yale University Press, 1984).

23. Jackson, "The Word Itself," *Discovering the Vernacular Landscape*, 7.

24. The resonance of classification as a structural schema in Enlightenment thought can also be found in Michel Foucault, *The Order of Things: An Archeology of the Human Sciences* (New York: Vintage, 1970), 125–65; and, as a regulating device in Foucault, Michel Foucault, "Docile Bodies," *Discipline and Punish: The Birth of the Prison* (New York: Pantheon Books, 1977), 135–69.

25. Dell Upton, "City as Material Culture," in *The Art and History of Historical Archeology*, ed. Elizabeth Yentsch and Mary C. Boudry (Boca Raton: CRC Press), 51–52.

26. Stilgoe, *Common Landscape of America*, 3.

27. Thomas Jefferson to Benjamin Rush, Monticello, 23 September 1800, *The Works of Thomas Jefferson*, ed. Paul L. Ford, 12 vols. (New York: G. P. Putnam's Sons, 1904–5), 9:147. For the classic analysis of the antiurban position, see Morton White and Lucia White, *The Intellectual versus the City: From Thomas Jefferson to Frank Lloyd Wright* (Oxford: Oxford University Press, 1962). The tensions and accommodations between pro- and antiurbanists are examined in Andrew

Lees, *Cities Perceived: Urban Society in European and American Thought, 1820–1940* (New York: Columbia University Press, 1985); Thomas Bender, *Toward an Urban Vision: Ideas and Institutions in Nineteenth Century America* (Baltimore: Johns Hopkins University Press, 1975); Janis P. Stout, *Sodoms in Eden: The City in American Fiction before 1860* (Westport, Conn.: Greenwood Press, 1976); Leo Marx, *The Machine in the Garden: Technology and the Pastoral Ideal in America* (New York: Oxford University Press, 1964); James Machor, *Pastoral Cities: Urban Ideals and the Symbolic Landscape of America* (Madison: University of Wisconsin Press, 1987).

28. See, e.g., Anders Stephanson, *Manifest Destiny: American Expansionism and the Empire of Right* (New York: Hill and Wang, 1995), 48–63.

29. The classic work on the subject of cities as spearheads in regional development which was advanced by Richard C. Wade has been augmented by other studies as well. *The Urban Frontier: Pioneer Life in Early Pittsburgh, Cincinnati, Lexington, Louisville, and St. Louis* (Chicago: University of Chicago Press, 1964). Leo Marx argues that "historians have long since discarded the beguiling [image] of a westward-moving frontier comprised of individual 'pioneers' like Daniel Boone or Natty Bumppo." "The Puzzle of Antiurbanism in Classic American Literature," in *Cities of the Mind: Images and Themes of the City in the Social Sciences*, ed. Lloyd Rodwin and Robert M. Hollister (New York: Plenum Press, 1984), 165. The urban geographer James E. Vance Jr. maintains that commercial cities drove American settlement, arguing that "wholesaling and cities came to America on the Mayflower and were just as much tools of pioneering as the axe and hoe." *The Merchant's World: The Geography of Wholesaling* (Englewood Cliffs, N.J.: Prentice-Hall, 1970), 10–11, 80–128. For a discussion of commercial colonization as a force in eighteenth-century settlement of Virginia's Shenandoah Valley and the interrelationships between agriculture, trade, and manufacturing in the commercialized landscape, see Robert D. Mitchell, *Commercialism and Frontier: Perspectives on the Early Shenandoah Valley* (Charlottesville: University Press of Virginia, 1977), 1–15, 161–229.

30. Joseph Blount, *An Anniversary Discourse, Delivered before the New-York Historical Society on Thursday, December 13, 1827* (New York: G. and C. Carvill, 1828), 49.

31. J. C. Myers, *Sketches on a Tour through the Northern and Eastern States, the Canadas and Nova Scotia* (Harrisonburg, Va.: J. H. Wartman, 1849), 112–13.

32. Basil Hall, *Travels in North American in the Years 1827–1828* (Edinburgh: Cadell, 1829), 153.

33. Noah Webster, *An American Dictionary of the English Language* (Springfield, Mass.: George and Charles Merriam, 1850), 586. The definitions were first presented in the 1828 edition.

34. Morton J. Horwitz, *The Transformation of American Law, 1780–1860* (Cambridge: Harvard University Press, 1977), 63–108.

35. Webster, *American Dictionary*, 586.

36. Clark, *Onondaga*, 2:73.
37. See Wyckoff, *Developer's Frontier*, 64–102. See also George Geddes, "The Erie Canal, Origin and History of the Measures That Led to Its Construction, Read before the Society, January 28, 1867," *Publications of the Buffalo Historical Society* (Buffalo: Bigelow Bros., 1880), 2:264.
38. Clinton quoted in William A. Bird, "New York State, Early Transportation, Read before the Society, January 29, 1866," *Publications of the Buffalo Historical Society* (Buffalo: Bigelow Bros., 1880), 2:20.
39. John L. Larson, *Internal Improvement: National Public Works and the Promise of Popular Government in the Early United States* (Chapel Hill: University of North Carolina Press, 2001), 75, quoting "Memorial of the Citizens of New-York, in Favor of a Canal Navigation between the Great Western Lakes and the Tide-waters of the Hudson," 21 February 1816, in New York, *Laws of the Canal*, 1:129. This kind of appeal percolated through the canal petitions. An 1816 petition in favor of the canal project argued that canals were "advantageous to towns and villages" by encouraging "country products for sale," "the commercial exchange of returning merchandize [sic]," "manufacturers," and an "increased economy and comfort of living"; it specifically prophesied that "great manufacturing establishments will spring up, agriculture will establish its granaries, and commerce its warehouses in all directions." "Memorial of the Citizens of New-York in Favor of a Canal Navigation," Ibid., 1:124. The economic reciprocity between towns and farms was not a new concept but one that had guided eighteenth-century settlement as well. See Mitchell, *Commercialism and Frontier;* James T. Lemon, *The Best Poor Man's Country: A Geographical Study of Early Southeastern Pennsylvania* (Baltimore: Johns Hopkins Press, 1972).
40. Larson, *Internal Improvement*, 75; Isaiah 35:1.
41. "Memorial of the Citizens of New-York in Favor of a Canal Navigation," 21 February 1816, *Laws of the State of New York, in Relation to the Erie and Champlain Canals*, 124.
42. Merwin S. Hawley, "Origin of the Erie Canal, Embracing a Synopsis of the Essays of the Hon. Jess Hawley, Published in 1807, Read before the Society, February 21, 1866," *Publications of the Buffalo Historical Society* (Buffalo: Bigelow Bros., 1880), 2:229.
43. Vance argues that improved transportation connections transformed localized "fundamental trading centers" into connected "distributing towns" and ultimately into important "unravelling points" that not only moved a variety of goods between the eastern entrepôts and western hinterlands but also became regionally dominant hubs in their own right. Vance, *Merchant's World*, 83–90. Muller's analysis of the "pioneer periphery," "specialized periphery," and "transitional periphery" focuses on balances between evolving local resource development and connections to external markets. Edward K. Muller, "Selective Urban Growth in the Middle Ohio Valley, 1800–1860," *Geographical Review* 66:2 (1976): 178–80.

44. O'Reilly, *Sketches of Rochester*, 168–242.
45. Ronald E. Shaw, *Canals for a Nation: The Canal Era in the United States, 1790–1860* (Lexington: University Press of Kentucky, 1990), 16.
46. Carol Sheriff, *The Artificial River: The Erie Canal and the Paradox of Progress, 1817–1862* (New York: Hill and Wang, 1996).
47. "Grand Canal Celebration," *Rochester Telegraph*, 1 November 1825.
48. Timothy C. Cheney, "Reminiscences of Syracuse," in *Annual Volume of the Onondaga Historical Association* (Syracuse: Dehler Press, 1914), 133–34; David Hoosack, *Memoir of De Witt Clinton: With an Appendix Containing Numerous Documents Illustrative of the Principal Events of His Life* (New York: J. Seymour, 1829), 354. The corollary was that towns that were left off the route suffered. In 1825 merchants in Manlius found "it is getting to be rather dull times with us here in consequence of our situation with respect to the canal, neither on it nor far enough from it not to feel that those immediately on its banks have a decided superiority over us in the transactions of most kinds of mercantile speculations." Five years later Azariah Smith complained, "we have found uniformly since the canal opened, a falling off in our business." Richard H. Schein, "A Historical Geography of Central New York: Patterns and Processes of Colonization on the New Military Tract, 1782–1820" (Ph.D. diss., Syracuse University, 1989), 294.
49. Benjamin DeWitt, "A Sketch of the Turnpike Roads in the State of New York," *Transactions of the Society for the Promotion of the Useful Arts* (Albany, 1807), 203–4, quoted in Schein, "Historical Geography of Central New York," 273.
50. As early as 1791, a Philadelphia publication advocated particular internal improvements that would "lay open the market of Philadelphia for the reception of the produce of the Genesee country." Bird, "New York State, Early Transportation," 2:19; Hawley, "Origin of the Erie Canal," 2:229, 255–56. In 1799 another account pointed out that the settlers found Canada "the most advantageous market." Bird, "New York State, Early Transportation," 2:19. See also Larson, *Internal Improvement*, 73–74.
51. Myers, *Sketches on a Tour*, 113.
52. Ibid.
53. Reginald Horsman, *Race and Manifest Destiny: The Origins of American Racial Anglo-Saxonism* (Cambridge: Harvard University Press, 1981). The seeds of this attitude were evident in the early colonial periods as well. See, e.g., William Cronon, *Changes in the Land: Indians, Colonists, and the Ecology of New England* (New York: Hill and Wang, 1983); Richard White, *The Middle Ground: Indians, Empires, and Republics in the Great Lakes Region, 1650–1815* (New York: Cambridge University Press, 1991).
54. Hoosack, *Memoir of De Witt Clinton*, 361.
55. "Observations on Navigable Canals," *America Medical and Philosophical Register* 1 (October 1810): 155.

56. Charles Caldwell, "Thoughts Regarding the Moral and Other Indirect Influences of Rail-Roads," *New-England Magazine* 2 (April 1832): 290, quoted in Richard D. Brown, "Emergence of Urban Society in Rural Massachusetts, 1760–1820," *Journal of American History* 61:1 (1974): 49.

57. Lees, *Cities Perceived*, 99, discussing Tappan, *Growth of Cities* (1855), 15, 24, 29.

58. Edwin Scrantom Diary, 7 August 1831, Rochester Public Library.

59. [M. D. Bacon?], "Great Cities," *Putnam's Monthly Magazine* 5 (March 1855): 254. That said, fears of moral corruption continued apace, both in literary circles and in the public arena. See White and White, *Intellectual versus the City*. See also "Town and Country," *Knickerbocker* 8 (November 1836): 537–39, quoted in Bayrd Still, *Urban America: A History with Documents* (Boston: Little, Brown, 1974), 195–96.

60. As Richard C. Wade argued: "nearly all the larger places owed their initial success to commerce. All sprang from it, and their great growth in the first half of the nineteenth century stemmed from its expansion." "The City in History—Some American Perspectives," in *Urban Life and Form*, ed. Werner Z. Hirsch (New York: Holt, Rinehart and Winston, 1963), 64. Roger Reifler seconded this premise, concluding that "for the antebellum period commercial activity, both interregional and especially intraregional trade, appears to be the driving force generating urbanization." "Nineteenth-Century Urbanization Patterns," 961. The examples of Rochester and Syracuse tend to support the commercial model of settlement proposed by James Vance that most interior cities "have not grown slowly out of the soil but quite the contrary, were sited in terms of the trading function whose long external links were clear and full before the first lot was platted." *Merchant's World*, 153.

61. Freeman Hunt, "The Good Merchant," *Worth and Wealth: A Collection of Maxims, Morals, and Miscellanies for Merchants and Men of Business* (New York: Stringer and Townsend, 1856), 57.

62. The theorist Ian Hacking has pointed out that "the forms of knowledge and power since the nineteenth century have served the bourgeoisie above all others" and noted the reciprocating loop in which "(1) a ruling class generates an ideology that suits its own interests; and (2) a new ideology, with new values, creates a niche for a new ruling class." "The Archaeology of Foucault," in *Foucault: A Critical Reader*, ed. David Couzens Hoy (New York: Basil Blackwell, 1986), 28, 27.

63. Hunt, "Mercantile Education," *Worth and Wealth*, 201.

64. David M. Scobey, *Empire City: The Making and Meaning of the New York City Landscape* (Philadelphia: Temple University Press, 2002), 33–42.

65. Brown, "Emergence of Urban Society," 48.

66. Ibid., 45.

67. For the impact of the merchant class on the new city of Rochester in particular, see, Paul Johnson, *A Shopkeeper's Millennium: Society and Revivals in*

Rochester, New York, 1815–1837 (New York: Hill and Wang, 1978). For new towns in the Middle West, see Timothy R. Mahoney, *Provincial Lives: Middle-Class Experience in the Antebellum Middle West* (New York: Cambridge University Press, 1999). For new towns in the South, see Lisa C. Tolbert, *Constructing Townscapes: Space and Society in Antebellum Tennessee* (Chapel Hill: University of North Carolina Press, 1999). For an examination of the depth of bourgeois influence on urban life, see Bushman, *Refinement of America;* and Blumin, *The Emergence of the Middle Class.* For a metropolitan perspective, see Scobey, *Empire City.* For the role of boosters in new towns in general, see Carl Abbott, *Boosters and Businessmen: Popular Economic Thought and Urban Growth in the Antebellum Middle West* (Westport, Conn.: Greenwood Press, 1981); and Daniel Boorstin, *The Americans: The National Experience* (New York: Vintage Books, 1965).

68. Vance, *Merchant's World,* 166, quoting Charles Dickens, *American Notes* (Greenwich, Conn.: Orenier Americana, 1961), 43, 41, 279.

69. Stone, "From New York to Niagara," 221.

70. Ibid.

TWO
Planning the Sorted City

1. William W. Campbell, *The Life and Writings of DeWitt Clinton* (Albany: Backer and Scribner, 1840), 176; Schein, "Historical Geography of Central New York," 309.

2. In 1817 Buffalo reportedly had one hundred houses, and by 1840 it held approximately 18,000 residents and Utica 12,800. Barber and Howe, *Historical Collections,* 148, 373; *Compendium of the Eleventh Census (1890),* 434–35.

3. McKelvey, *Rochester: The Water-Power City,* 24–34.

4. Letter from Nathaniel Rochester to Charles Carroll, 13 January 1811, quoted in "Documents of Early Rochester: Some Recent Acquisitions of the Society," *Rochester Historical Society Genesee County Scrapbook* 3 (1957): 13.

5. Vance, *Merchant's World,* 8, 48–51; "The American City: Workshop for a National Culture," in *Contemporary Metropolitan America,* vol. 1: *Cities of the Nations Historic Metropolitan Core,* ed. John S. Adams (Cambridge: Ballinger, 1976), 16–19; Muller, "Selective Urban Growth," 185–92; William Cronon, *Nature's Metropolis: Chicago and the Great West* (New York: W. W. Norton, 1981), 278–309; Sherry Olson, "Baltimore Imitates the Spider," *Annals of the Association of American Geographers* 69:4 (December 1979): 557–74; "Urban Metabolism and Morphogensis," *Urban Geography* 3:2 (April–June 1982): 87–109; Neil L. Shumsky, "Introduction," in *Urbanization and the Growth of Cities* (New York: Garland, 1996), xv–xvi.

6. The dominance of commercial values is noted also in Bushman, *Refinement of America,* 154; Conzen, "Morphology of Nineteenth-Century Cities," 127.

7. E.g., an early ordinance directed that wagons may not "stop in the square formed by the intersection." *A Directory for the Village of Rochester Containing the Names, Residence and Occupation of All Male Inhabitants... To Which Is Added a Sketch of the History of the Village* (Rochester: Elisha Ely, 1827), 102.

8. Daniel Bluestone, *Constructing Chicago* (New Haven: Yale University Press, 1991), 16, fig. 4, 154–58.

9. Conzen, "Morphology of Nineteenth-Century Cities"; Edward T. Price, "The Central Courthouse Square in the American County Seat," *Geographical Review* 58:1 (1968): 29–60; Richard Pare, *Court House: A Photographic Document* (New York: Horizon Press, 1978); Tolbert, *Constructing Townscapes;* John W. Reps, *Cities of the American West: A History of Frontier Urban Planning* (Princeton: Princeton University Press, 1979); *The Making of Urban America: A History of City Planning in the United States* (Princeton: Princeton University Press, 1965); *Town Planning in Frontier America* (Princeton: Princeton University Press, 1969); Wilber Zelinsky, "The Pennsylvania Town: An Overdue Geographical Account," *Geographical Review* 67:2 (April 1977): 127–47; Joseph S. Wood, *The New England Village* (Baltimore: Johns Hopkins University Press, 1997).

10. Marcus Christian Hand, *From a Forest to a City: Personal Reminiscences of Syracuse* (Syracuse: Masters and Stone, 1889), 4. Conzen points out that space for moving people and goods took up more space in American plans than in European ones, from which he deduces that Europeans were more concerned with aesthetic effects of their streetscape and the Americans with efficiency. "Morphology of Nineteenth-Century Cities," 120. This overlooks the idea that broad convenient streets could be thought of as a beautiful component to a cityscape.

11. See also Wyckoff, *Developer's Frontier*, 14–44.

12. Bird, "New York State, Early Transportation," 2:29; Hawley, "Origin of the Erie Canal," 2:245, 252–54; Geddes, "Erie Canal Origin," 2:276–78.

13. Thomas Cooper, *A Ride to Niagara in 1809* (Rochester: George P. Humphrey, 1915), frontis map.

14. O'Reilly, *Sketches of Rochester*, 247–49.

15. McKelvey, "Colonel Nathaniel Rochester," 8–9. Even before the bridge was completed in 1812, it was causing concern in towns to the south. A Geneseo developer rued: "I wish the tract of 100 acres could be purchased of the Maryland gentlemen. The Bridge and Mill seat render it very valuable indeed." Turner, *Phelps and Gorham's Purchase*, 587, quoted in McKelvey, *Rochester: The Water-Power City*, 35. For the economic influence of New York turnpikes, see Michael N. Gimigliano, "Experiences along the Cherry Valley Turnpike: The Education of a Traveller" (Ph.D. diss., Syracuse University, 1979). Since the bridge would benefit both Genesee and Ontario counties, on the west and east banks of the river, respectively, the cost of "an excellent bridge of uncommon strength" was shared by both. DeWitt Clinton, "Private Canal Journal, 1810," in *The Life and Writings of DeWitt Clinton*, ed. William Campbell (New York: Baker and Scrib-

ner, 1849), 112. See also William Peck et. al., *Landmarks of Monroe County, New York* (Boston: Boston History Co., 1895), 109.

16. Price, "Central Courthouse Square."

17. David Hamer, *New Towns in the New World: Images and Perception of the Nineteenth-Century Frontier* (New York: Columbia University Press, 1990), 40–64.

18. John Fowler, *Journal of a Tour in the State of New York in the Year 1830 with Remarks on Agriculture in Those Parts Most Eligible for Settlers* (1831; rpt., New York: Augustus M. Kelley 1970), 103. Those representations came from the published literature Fowler brought with him, including Spafford's *Gazetteer of New York*, which Fowler in turn plagiarized in his own published account. The New England diarist Thomas Woodcock also used published travel accounts to prepare for his 1836 journey to Niagara Falls; his unpublished diary includes selections from the published travelers Basil Hall and Fanny Kemble. Thomas Swann Woodcock, "Trip from New York to Niagara, 1836," New York Public Library. Moses Cleveland's 1831 journal also borrowed phrases from Vandewater's *Tourist Guide*. "Journal of a Tour from Riverhead, Long Island, to the Falls of Niagara in June 1831, by Moses C. Cleveland," ed. N. R. Howell, *New York History* 27 (July 1946): 356. Travel accounts were a popular genre that received reviews in literary magazines; see, e.g., review of Charles Yell's *Travels in North America*, in *American Review* (October 1845): 403–7; and review of Basil Hall's *Travels in North America*, in *American Monthly Magazine* (1820): 532–40. The ubiquity of the travel writing genre resulted in the lampoon "The Science of Criticism Systematized, or, The Art of Reviewing Made Easy," *American Monthly Magazine* (1829): 329.

19. James E. Vance Jr., *The North American Railroad: Its Origin, Evolution, and Geography* (Baltimore: Johns Hopkins University Press, 1995), xii.

20. Horatio Gates Spafford, *A Gazetteer of the State of New-York: Embracing an Ample Survey and Description of Its Counties, Towns, Cities, Villages* (Albany: D. B. Packard, 1824), 462.

21. The price of thirty and fifty dollars per quarter-acre lot was in keeping with other town founders in central New York: quarter-acre lots in recently platted Ithaca, Ovid, and Jefferson were selling for twenty to fifty dollars in 1810. Schein, "Historical Geography of Central New York," 317.

22. "Documents of Early Rochester," *Rochester Historical Society Genesee County Scrapbook*, 12–13.

23. Deeply in debt by 1818, Forman and his associates, Ebenezer Willson and William Taylor, sold out to Daniel Kellog and William H. Sabin, Forman's brother-in-law and brother's law partner in Onondaga Hollow, respectively. They in turn sold it to the shipping magnate Henry Eckford of New York City for twenty-five thousand dollars, who sold it to the Syracuse Company, which kept Forman on as their land agent. Galpin, "Genesis of Syracuse," *New York History*,

NOTES TO PAGES 27–30

23. Letter from Henry Eckford to Joshua Forman, 19 December 1821, New York State Archives.

24. Daniel Drake (1815), quoted in Still, *Urban America*, 63. "The streetgrid historically has been a common symbol . . . for rational urban life," argues cultural geographer Paul Groth in "Streetgrids as Frameworks for Urban Variety," *Harvard Architecture Review* 2 (Spring 1981): 69.

25. Emerson continues, "The Country, on the contrary, offers an unbroken horizon, the monotony of an endless road, of vast uniform plains, of distant mountains . . . the eye is invited ever to the horizon and the clouds. It is the school of Reason." Cited in Lees, *Cities Perceived*, 94, and Cowan, *City of the West*, 61–62.

26. Francis Baily, *Journal of a Tour in the Unsettled Parts of North America*, 105, quoted in Reps, *Making of Urban America*, 294.

27. Estwick Evans, *A Pedestrious Tour of Four Thousand Miles through the Western States and Territories during the Winter and Spring of 1819* (Concord, N.H.: Joseph C. Spear, 1819), reprinted in Reuben Gold Thwaites, *Early Western Travels, 1748–1846* (Cleveland: Arthur H. Clark, 1904), 8:117.

28. Reps, *Making of Urban America*, 294–302. Peter Marcuse, "The Grid as City Plan," *Planning Perspectives* 2 (1987): 287–310; Blackmar, *Manhattan for Rent*, 94–100; Harold M. Mayer and Richard C. Wade, *Chicago: Growth of a Metropolis* (Chicago: University of Chicago Press, 1969), 14–26.

29. See Martyn J. Bowden, Edward K. Muller, John A. Radford, and Peter Rees, "Prospectus for The Mercantile City in America: A Geographical Perspective," MS. See also Blackmar, *Manhattan for Rent*, 77–87.

30. Muller, "Baltimore," in "Prospectus for The Mercantile City in America."

31. Freeman Hunt, *Letters about the Hudson River and Its Vicinity, Written in 1835–1837* (New York: Freeman Hunt, 1837), 44.

32. John Woodworth, *Reminiscences of Troy, from Its Settlement in 1790 to 1807, with Remarks on Its Commerce, Enterprise, Improvements, State of Political Parties, and Sketches of Individual Character* (Albany: J. Munsell, 1860), 24; David Buel, *Troy for Fifty Years: A Lecture Delivered before the Young Men's Association of Troy on December 21, 1840* (New York: N. Tuttle, 1841), 13.

33. Esperanza was later physically and officially reconfigured as Athens, New York. Reps, *Making of Urban America*, 352–55.

34. "Commissioners' Remarks," in William Bridges, *Map of the City of New York and Island of Manhattan* (New York, 1811), 24–25, quoted in Reps, *Making of Urban America*, 297.

35. John Love, *Geodaesia: or, The Art of Surveying and Measuring Land Made Easy . . . Also, to Layout New Lands in America, or Elsewhere* (New York: Samuel Campbell, 1796), preface. For other English language surveying books, see, e.g.,

A. Nesbit, *A Treatise on Practical Mensuration in Eight Parts* (London: Longman, Hurst, Rees, Orme, and Brown, 1816); William Davis, *A Complete Treatise on Land Surveying by the Chain, Cross, and Offset Staffs* (London, 1798); J. Cotes, *The Surveyor's Guide; or a Treatise on Practical Land Surveying* (London: B. and R. Crosby, 1814).

36. Reps, *Making of Urban America*, 317–22, 488–90.

37. Letter from Charles Carroll to Nathaniel Rochester, 17 August 1811, quoted in "Documents of Early Rochester," *Rochester Historical Society Genesee County Scrapbook*, 12.

38. Peck, *History of Monroe County*, 73; Osgood, "Rochester," 65–67.

39. As the urban geographer James Vance has shown, new cities in "pioneer regions" required economic assistance from a second sector as well, "from trading, either in the sense of securing and distributing manufactures or in collecting and forwarding the natural produce." Vance, *Merchant's World*, 96–97, 153, 48–51. Vance, "American City," 16–19. See also Olson, "Metabolism and Morphogenesis"; and "Baltimore Imitates the Spider."

40. John I. Bigsby, "Remarks on the Environs of Carthage Bridge," *American Journal of Science and Arts* (1820): 252.

41. O'Reilly, *Sketches of Rochester*, 263.

42. "The Social Statistics of Cities," *The Miscellaneous Documents* (1880 census), 13:xxx.

43. Letter from Charles Carroll to Nathaniel Rochester, 11 August 1811, quoted in "Documents of Early Rochester," *Rochester Historical Society Genesee County Scrapbook*, 12. W. H. McIntosh, *History of Monroe County, New York; With Illustrations Descriptive of Its Scenery, Palatial Residences, Public Buildings, Fine Blocks and Important Manufactories, 1788–1877* (Philadelphia: Everts, Ensign and Evers, 1877), 73; Osgood, "Rochester," 65–67; Hamlet Scrantom, "Scrantom Letters on the Beginnings of Rochester," *Rochester Historical Society Publication Fund Series* 7 (1928): 174.

44. Richard Longstreth, *The Buildings of Main Street: A Guide to American Commercial Architecture* (Washington, D.C.: Preservation Press, 1987), 24–29. A traveler's observation of Troy, New York, rang true for many of the new cities in New York: "the houses are very neat and numerous; almost every house contains a shop." Duke de la Rochfoucault-Liancourt (1795), quoted in Arthur J. Weise, *Troy's One Hundred Years, 1789–1889* (Troy: W. H. Young, 1891), 44.

45. Nathaniel Rochester Cash Book, Rochester Public Library. Scrantom, "Scrantom Letters," 174. A similar pattern could be found in Manlius in central New York, where the quarter-acre town lots sold for approximately thirty dollars apiece, except for two lots at the intersection of the turnpike which sold for five hundred dollars each. Schein, "Historical Geography of Central New York," 271–72.

46. James E. Vance Jr., "Focus on Downtown," in *Internal Structure of the*

NOTES TO PAGES 34–39

City: Readings on Space and Environment, ed. Larry S. Bourne (New York: Oxford University Press, 1971), 115.

47. History of Monroe County, 73; Osgood, "Rochester," 65–67; Scrantom, "Scrantom Letters," 174; Alphonso A. Hopkins, The Powers Fire-Proof Commercial and Fine Arts Buildings (Rochester: E. R. Andrews, 1883); James Matthews, A Description of Powers' Commercial Fire-Proof Building, with the Most Important Statistics Connected with Their Construction, and Enumeration of the Most Interesting Objects to Be Seen from the Tower (Rochester: E. R. Andrews, 1872).

48. Nathaniel Rochester Cash Book, Rochester Public Library. As early as 1811, the merchant miller Shuball Hovey contemplated taking a ten-year lease on Rochester's mill site and buying the first two lots on the southwest corner of Buffalo and Mill for a tavern and store.

49. Letter to Charles S. Baker from Abelard Reynolds, 10 April 1872, "1873 Rochester Time Capsule," <www.rmsc.org/capsule/2000%201%20329a.htm>.

50. "Rochester's History" (1884 reminiscence), H. H. H. Barton Scrapbook, 1:16, 1, Rochester Public Library Local History Division.

51. Mahoney, Provincial Lives, 168–212.

52. Clark, Onondaga, 2:70.

53. Ibid.

54. William Cooper, A Guide in The Wilderness or the History of the First Settlements in the Western Counties of New York with Useful Instructions to Future Settlers (1810; rpt., Freeport: Books for Libraries Press, 1970), 7–8, 200. Later remembered as the father of James Fenimore Cooper, in his own day William Cooper was well known as the founder of Cooperstown and developer of forty thousand acres of land in New York State. It is unclear whether the founders of Rochester or Syracuse actually knew of Cooper's posthumously published advice or the specifics of his city-building principles, but many of their own decisions reflected similar strategies, indicating the cultural currency of those ideas.

55. Ibid., 17–20.

56. Pred, Urban Growth and City-Systems, 1.

57. Cooper, Guide in the Wilderness, 19.

58. Ibid., 17–18.

59. Perhaps Cooper's ideas were derived from his gridded hometown of Burlington, New Jersey, which was not far from the famous grid of Philadelphia. In 1786 he laid out his own eponymous village in an orthogonal plan of small lots ranging from 25 to 40 feet wide and 150 feet deep.

60. Letter from J. Townsend to H. Baldwin, 17 February 1834, Townsend Family Business Records, New York Public Library.

61. Letter from Charles Carroll to Nathaniel Rochester, 11 August 1811, quoted in "Documents of Early Rochester," Rochester Historical Society Genesee County Scrapbook, 12. "Rochester's History [1884]," Barton Scrapbook, 1:16.

62. [Johnson Verplanck?], Our Travels, Statistical, Geographical, Mineralogi-

cal, Geological, Historical, Political, and Quizzical: A Knickerbocker Tour of New York State, 1822, ed. Louis Leonard Tucker (Albany: New York State Library, 1968), 94–95.

63. See, e.g., Blumin, The Emergence of the Middle Class; Bushman, Refinement of America; Johnson, Shopkeeper's Millennium; Mahoney, Provincial Lives; Don Harrison Doyle, The Social Order of a Frontier Community: Jacksonville, Illinois, 1825–70 (Urbana: University of Illinois Press, 1978).

64. Cooper, Guide in the Wilderness, 19.

65. James E. Vance Jr., The Continuing City: Urban Morphology in Western Civilization (Baltimore: Johns Hopkins University Press, 1990), 13; Conzen, "Morphology in Nineteenth-Century Cities," 125.

66. Wyckoff, Developer's Frontier, 15.

67. Letter from Nathaniel Rochester to Charles Carroll, 13 January 1811, quoted in "Documents of Early Rochester," Rochester Historical Society Genesee County Scrapbook, 13.

68. Ibid., 11–12.

THREE
Building the Sorted City

1. Mrs. Basil [Margaret Hunter] Hall, The Aristocratic Journey: Being the Outspoken Letters of Mrs. Basil Hall Written during a Fourteen Months' Sojourn in America, 1827–1828 (New York: G. P. Putnam's Sons, 1931), 53; Hall, Travels in North America, 152–67; Forty Etchings: From Sketches Made with Camera Lucida in North America, in 1827 and 1828 (London: Cadell, 1830), n.p.

2. Hall, Travels in North America, 161.

3. Ibid., 163–64.

4. Ibid., 160–61.

5. Ibid., 161.

6. Esther Maria Ward Chapin, "Diary," 30 January 1821, in "Diaries and Letters, 1815–1823," Rochester Public Library.

7. John Howison, Sketches of Upper Canada, Domestic, Local, and Characteristic: To Which Are Added Practical Details for the Information of Emigrants of Every Class; and Some Recollections of the United States of America (London: G. and W. B. Whittaker, 1821), 285.

8. The bridge was replaced one more time during the 1850s; it still stands today.

9. A Directory for the Village of Rochester (1827). The initial occupancy and construction in a new city was "the most enduring physical aspect of the city." Groth, "Streetgrids as Frameworks," 72.

10. Caroll E. Smith, Pioneer Times in the Onondaga Country (Syracuse: C. W. Bardeen, 1904), 233, 240, see also 265; Cheney, Reminiscences of Syracuse, 118 ff.;

Hand, *Forest to City*, 103; "Reminiscences of Syracuse," *Syracuse Standard*, 19 June 1867, Onondaga Historical Association (OHA) Clippings file; Sterry, "Some Reminiscences," 1 July 1883, OHA Reminiscences file.

11. Vance, *Merchant's World*, 17, 21–33, 96–97.

12. *Directory for the Village of Rochester* (1827), n.p.

13. "Some Reminiscences," *Syracuse Times*, 1 July 1883, OHA Reminiscences file.

14. Cheney, "Reminiscences," 97.

15. Ibid., 134; Clark, *Onondaga*, 2:92.

16. Bigsby, "Remarks on the Environs of Carthage Bridge," 252.

17. "To the Public," *Rochester Telegraph*, 3 November 1818. Nathaniel Rochester began petitioning the state legislature as early as 1817 to create a new county. For several years the legislature kept tabling the petition, much to the ire of Rochester residents, who accused the body of "legislative intrigue, bargain and sale, corruption and management." Rochesterians accused "the lofty aristocrats of the surrounding country, [who] may volunteer in crusades to Albany, and with superior influence and superior knowledge of men in the state, succeed year after year, by defeating each application." Ibid. The urban geographer James Vance has argued that established cities opposed new towns because of the threat to their wholesaling dominance in the region. *Merchant's World*, 78–79.

18. McKelvey, *Rochester: The Water-Power City*, 73–74.

19. Chapin Diary, 30 January, 1821; emphasis omitted.

20. Rochester Laws of 1822, chapter 192, cited in Edward Foreman, ed., *Centennial History of Rochester*, 2 vols. (Rochester: Rochester Public Library, 1931), 2:6.

21. Letter from Matthew Brown to Nathaniel Rochester, 15 May 1817, reprinted in "Village Letters," *Rochester Historical Society*, 7; Johnson, *Shopkeeper's Millennium*, 63–66; Blake McKelvey, "The Physical Growth of Rochester," *Rochester History* 8:4 (October 1951): 3.

22. The civic district persisted as an adjunct space on the direct edge of the commercial district. In 1851 a new courthouse was constructed on the site. In 1874 a separate city hall was built on the site of the First Presbyterian Church, and in 1894 the old courthouse was replaced with a Beaux Arts edifice. Both institutions still stand today.

23. "Map of Rochester" (Rochester: Silas Cornell, 1839), Rochester Public Library.

24. Amos Granger sweetened the deal by offering to construct a fireproof clerk's office and to contribute a thousand dollars toward the courthouse enterprise. During the 1840s Granger again tried to draw the courthouse to Syracuse with his offer to build the courthouse in the center for twenty thousand dollars plus the old courthouse and lot. This offer was also refused. Smith, *Pioneer Times in Onondaga County*, 342–44.

25. Much like Genesee and Ontario counties, Onondaga County later lost

territory as the Military Tract became more settled. Cayuga (1799), Cortland (1808), and Oswego (1816) counties were shaved from the original borders. The debate over the location of the courthouse is recounted in several sources, including Cheney, "Reminiscences," 152–58.

26. "Reminiscences of Syracuse," *Syracuse Daily Journal*, 30 July 1845, in Tredwell Scrapbook, 381, OHA.

27. James Silk Buckingham, *America, Historical, Statistic, and Descriptive*, 2 vols. (New York: Harper and Bros., 1841), 2:247.

28. "Early Syracuse—Dr. A. R. Morgan Indulges in Some Reminiscences," OHA Reminiscences file.

29. Smith, *Pioneer Times*, 311.

30. Hand, *Forest to a City*, 199.

31. Stone, "From New York to Niagara," 221.

32. "Cheney to the Board of Supervisors on Onondaga County," Timothy C. Cheney Papers, n.d., OHA.

33. Ibid.

34. Cheney, "Reminiscences," 157.

35. Ibid. White later designed similar courthouses in brick for Chemung County and Jefferson County in New York. Elinore Taylor Horning, *The Man Who Changed the Face of Syracuse: Horatio Nelson White* (Mexico, N.Y.: Elinore T. Horning, 1988), 17–19.

36. Buckingham, *America*, 2:248.

37. Nineteenth-century mercantile cities often began as "depots of staple collection," where wholesalers operated with the dual purpose of collecting and dispersing local products. Vance, *Merchant's World*, 154.

38. Letter from Charles Carroll to Nathaniel Rochester, 17 August 1811, quoted in "Documents of Early Rochester," *Rochester Historical Society Genesee County Scrapbook*, 12.

39. Letter from Charles Carroll to Nathaniel Rochester, 9 November 1817, reprinted in "Village Letters," *Rochester Historical Society*, 11.

40. Ibid.

41. William Brown, *America: Four Years Residence in the United States and Canada: Giving a Full and Fair Description of the Country, as It Really Is* (Leeds: William Brown, 1849), 9–12.

42. Barber and Howe, *Historical Collections*, 267.

43. A term derived from David Nye, *American Technological Sublime* (Cambridge: MIT Press, 1994).

44. A. Duttenhofer, *Bereisung der Vereiningten Staaten von Nordamerika, mit besonderer Hinsicht auf den Erie-Canal* (Stuttgart: C. W. Löflund, 1835), 39. (Trans. Kai Gutschow.)

45. N. P. Willis, *American Scenery, or, Land, Lake, and River: Illustrations of Transatlantic Nature*, illus. W. H. Bartlett (London: George Virtue, 1840), 80.

NOTES TO PAGES 56–59

46. Charles A. Dana, *The United States Illustrated; in Views of Town and Country* (New York: Hermann J. Meyer, 1854), 59.

47. Frederick von Raumer, *America and the American People*, trans. William W. Turner (New York: J. and H. G. Lagley, 1846), 457.

48. Michael Hutchinson Jenks, "Notes on a Tour through the Western Parts of the State of New York, May 1828," 47, Department of Rare Books and Special Collections, University of Rochester Libraries.

49. Trollope, *Domestic Manners of the Americans*, 333; Frederika von Bremer, *The Homes of the New World: Impressions of America*, 2 vols., trans. Mary Howitt (New York: Harper and Bros., 1853), 1:580.

50. Thomas Rolph, *A Brief Account, Together with Observations, Made during a Visit in the West Indies, and a Tour through the United States of America, in Parts of the Years 1832–3* (Dundas, Ont.: Heyworth Hacksack, 1836), 95. For openness generally, see Hamer, *New Towns in the New World*, 52–55. Even if the traveler were not anxious to learn of the city's manufacturing genius, he found it difficult to avoid the milling and manufacturing scene. "I was shewn a very ingenious and newly-invented apparatus for making nails, together with various other pieces of curious mechanism," John Howison wrote, "which I had neither time nor inclination to inspect minutely." *Sketches of Upper Canada*, 286. Of course, not every miller was willing to explain his inventions. The Rochester miller Harding became close-mouthed when one traveler asked for more information in order to build a mill in Virginia. Emily E. E. Whyte, "Diary of Emily Esther Elizabeth Whyte, 1843," 38, Rochester Public Library.

51. Whyte, "Diary," 38. Some travelers carried letters of introduction, such as Basil and Margaret Hall's recommendation from Governor Clinton which gave them access to the city's leaders and thus its sites. Hall, *Aristocratic Journey*, 49.

52. Buckingham, *America*, 2:196, 189, 247–48.

53. Edwin Perry Clapp, "Travel, Trade and Transportation of the Pioneers" MS (1907), 161–62, Rochester Public Library,

54. "Reminiscences of Rochester," Edwin Scrantom Scrapbook, 2 vols., 1:53–54, Rochester Public Library. The son of Charles Carroll, one of the original founders for whom Carroll Street was named, sued the city over payments for the rights to the embankment that he had constructed. The angered trustees retaliated by revoking the name of Carroll Street.

55. Parker, *Story Historical*, 164.

56. The New York gazetteer Horatio Spafford noted in 1813, and again in 1824, that the salt trade employed "a very large proportion of the inhabitants, who are necessarily collected into clusters around the various works." See, e.g., Spafford, *Gazetteer of New York* (1824), 461.

57. Fowler, *Journal of a Tour*, 87; Buckingham, *America*, 2:247.

58. French, *Historical and Statistical Gazetteer*, 481.

59. Barber and Howe, *Historical Collections*, 398.

60. Archibald M. Maxwell, *A Run through the United States, during the Autumn of 1840*, 2 vols. (London: Henry Colburn, 1841), 1:235. See also George Combe, *Notes of the United States of North America, during a Phrenological Visit in 1838–39–40.*, 3 vols. (Edinburgh: Maclachlin, Stewart, 1841), 2:323.

61. Hall, *Notes of a Tour in 1829*, 27.

62. Salt humor is today a lost genre, but it was in full force during the nineteenth century. In "Salt City," they said, corned beef roamed the streets, and pickled perch swam in the stream. In Salt City it never rains; it brines. When a lady from Syracuse kisses her friend from Watertown, she says "Why, how fresh you are," and, when the Watertown lady returns the kiss, she says to herself, "I'm sorry I can't return the complement," and asks for a drink of water. In another satire with a biblical twist, *Vanity Fair* alluded both to Syracuse's rapid urbanization and its salt economy, two factors inextricably bound together: Lot's wife turned into a pillar of salt when she looked back over one of the canal bridges. "Mrs. Lot is no more, but there are lots of Lots left in Syracuse, and corner ones are quite valuable." "Salty Syracuse," *Syracuse Daily Journal*, 16 January 1873, clipping in OHA Travelers Impressions file; "Vanity Fair's Easy Lessons in Geography—No. 1," *Vanity Fair*, 12 October 1861, 177.

63. O'Reilly, *Sketches of Rochester*, 225.

64. Ibid., 224–25; Shaw, *Erie Water West*, 186–87; McKelvey, *Rochester: The Water-Power City*, 147.

65. For an extensive social analysis of public displays and processions, see Susan G. Davis, *Parades and Power: Street Theater in Nineteenth-Century Philadelphia* (Philadelphia: Temple University Press, 1986); Sean Wilentz, *Chants Democratic: New York City and the Rise of the American Working Class, 1788–1850* (New York: Oxford University Press, 1984), 87–97. For an architectural and urban focus on parades, see Spiro Kostof, *The City Assembled: The Elements of Urban Form through History* (Boston: Bullfinch Press, 1992), 194–97.

66. "Grand Canal Celebration," *Rochester Telegraph*, 1 November 1825.

67. Ibid.

68. Johnson, *Shopkeeper's Millennium*.

69. "Grand Canal Celebration," *Rochester Telegraph*, 1 November 1825. An ailing Nathaniel Rochester was too sick to attend, but sent his toast by proxy: "the greatest publick [sic] work in America, if not in the world. A principal link in the chain that binds the Union of the States."

70. Ibid.

71. Blake McKelvey, "Rochester and the Erie Canal," *Rochester History* 11:3–4 (July 1949): 6.

72. O'Reilly, *Sketches of Rochester*, 224.

73. Hall, *Travels in North America*, 162.

74. [Edwin Scrantom], "And This Was Rochester! Excerpts from the Old Cit-

NOTES TO PAGES 63–67

izen Letters of Edwin Scrantom," *Rochester History* 4 (January 1942): 9. See also Hall, *Travels in North America*, 160.

75. Stephen Davis, *Notes of a Tour in America in 1832 and 1833* (Edinburgh: Waugh and Innes, 1833), 46–47.

FOUR
Refining the Sorted City

1. Barber and Howe, *Historical Collections*, 267, 394; French, *Historical and Statistical Gazetteer*, 109–10, 406, 490.
2. Barber and Howe, *Historical Collections*, 4.
3. Ibid., 395. The built environment similarly compelled another traveler to conclude that, despite the legal status of Syracuse as a village, it was "in reality a very large and city-looking town." Miss Leslie, "Western New York: A Slight Sketch," *Lady's Book* 31 (November 1845): 185.
4. The list of bankruptcies and financial reversals included Abelard Reynolds, William Atkinson, Hervey Ely, James Livingston, and Selah Matthews, all members of the Rochester elite. See McKelvey, *Rochester: The Water-Power City*, 220–21. For a discussion of the social currency of entrepreneurship in nineteenth-century cities, see Peter G. Goheen, "Industrialization and the Growth of Cities in Nineteenth-Century America," *American Studies* 14:1 (1973): 58–59.
5. Leslie, "Western New York," 185.
6. "Opinions of Rochester," 1840s newspaper clipping in Augustus Raymond Scrapbook, 56, Rochester Public Library.
7. R. J. Vandewater, *The Tourist, or Pocket Manual for Travellers on the Hudson River, the Western Canal, and Stage Road, to Niagara Falls*, 2d ed. (New York: Ludwig and Tolefree, 1831), 53.
8. Ibid., 48. The same phrase, "New York in miniature," was used by Moses Cleveland during his 1831 foray down the Erie Canal. "Journal of a Tour from Riverhead, Long Island to the Falls of Niagara in June 1831, by Moses C. Cleveland," ed. N. R. Howell, *New York History* 27 (July 1946): 356.
9. "Early Rochester," 1884 clipping in H. H. H. Barton Scrapbook, 1:50.
10. *Syracuse Business Directory* (Syracuse: 1853), xx.
11. The author was likely either Daniel Wadsworth of Hartford or the editor Benjamin Silliman of New Haven. "Architecture in the United States," *American Journal of Science and Arts* 17:1 (January 1830): 101.
12. "Architecture in the United States," *American Journal of Science and Arts* 17:2 (January 1830): 253, 254–57. See also "Architecture in the United States," *American Journal of Science and Arts* 18 (July 1830): 11–27, 212–37. "Architecture in the United States," *American Journal of Science and Arts* 24:2 (July 1833): 257–63.

13. "Architecture in the United States," 17:2:259.
14. Ibid., 17:1:108.
15. Ibid., 17:1:101.
16. Ibid., 17:1:102.
17. "Petition by A. N. Van Patten to build an office," 8 March 1834, Canal Board Records, New York State Archives.
18. Of course, appearances could also lie. A disgruntled resident who had learned that appearances were not what they seemed warned his fiancée not to too expect too much of Rochester: "One would suppose from the promising appearance of this Village, that the people were flourishing—But I assure you Connecticut maxims & habits are reversed, & nothing is more easy than to build a fine house or stores & dash into trade & speculation, upon *other peoples* money." "Letter from Joseph Spencer to Elisabeth Selden, July 9, 1818," reprinted in "Village Letters," *Rochester Historical Society*, 11–12. Bushman extensively discusses the legibility of the city and the importance of aesthetic built environments that "could be evaluated for their taste and beauty." Bushman, *Refinement of America*, 139
19. "Plant Trees, Plea in 1829," "Syracuse History Newspaper Clippings Scrapbook," 23, OHA.
20. Dwight H. Bruce, *Onondaga's Centennial: Gleanings of a Century* (Boston: Boston History Co., 1896), 440; *Memorial History of Syracuse, New York, From Its Settlement to the Present Time* (Syracuse: H. P. Smith, 1891), 143.
21. Ibid.; Cheney, "Reminiscences," 121; "Principal Blocks in 1847 Merely Memories Now," *Syracuse Journal*, 14 April 1925; "In Business Fifty Years," clipping in OHA Reminiscences file, n.d.
22. "The Working-Man's Dwelling," *New Genesee Farmer* (August 1841): 127.
23. Catherine E. Beecher and Harriet Beecher Stowe, *The American Women's Home or, Principles of Domestic Science* (1869; rpt., Hartford: Harriet Beecher Stowe Center, 1975), 84; Catherine E. Beecher, *A Treatise on Domestic Economy for the Use of Young Ladies at Home and at School* (New York: Harper, 1846).
24. Eliza Steele, *A Summer Journey in the West* (New York: John S. Taylor, 1841), 36–37.
25. Buel, *Troy for Fifty Years*, 34–35.
26. James Jackson Jarves, *The Art-Idea: Sculpture, Painting and Architecture in America* (Boston: Houghton Mifflin, 1864), 264–65. Jarves also praised merchants for hiring architects, a professional capable of "designing an edifice which is to distinguish his business." Ibid.
27. Ibid., 267.
28. Boorstin, *National Experience*, 142.
29. Ibid., quoting Anthony Trollope.
30. Barber and Howe, *Historical Collections*, 395.
31. Howison, *Sketches of Upper Canada*, 303.

32. *Sarmiento's Travels in the United States in 1847*, trans. Michael A. Rockland (Princeton: Princeton University Press, 1970), 129.

33. [Hawthorne], "Sketches from Memory," 400.

34. According to an 1815 dictionary, a *tavern* was "a house where wine is sold, and drinkers are entertained; a house where travellers are entertained." An *inn* was "a house of entertainment for travellers; a house where students are boarded and taught." A *hotel* was characterized by its tenor, not its services; a hotel was "a genteel inn." John Walker, *A Critical Pronouncing Dictionary, and Expositor of the English Language* (Philadelphia: Kimber and Richardson, 1815), 533, 303, 470. A large hotel would include in it staff, clerks, bartenders, cooks, waiters, scullery maids, chambermaids, laundresses, maintenance men, and possibly a livery service. Although the terms *inn*, *tavern*, and *hotel* could be used interchangeably, around midcentury *hotel* typically connoted a more genteel, fuller-service institution.

35. An 1834 advertisement in Foreman, *Centennial History of Rochester*, 2:155.

36. John T. Roberts, "Business Men of the Village," *Publications of the Onondaga Historical Association* 1 (1910): 31.

37. Quoted in Franklin Henry Chase, *Syracuse and Its Environs: A History*, 3 vols. (New York: Lewis, 1924), 305.

38. Buckingham, *America*, 2:246.

39. Hamer, *New Towns in the New World*, 43–44.

40. "Journal of a Trip to Niagara, 1822," Department of Rare Books and Special Collections, University of Rochester Libraries.

41. [Verplanck?], *Knickerbocker Tour of New York State, 1822*, 99.

42. Quoted in Chase, *Syracuse and Its Environs*, 306; Hand, *From a Forest to a City*, 33.

43. Letter from General Granger to J & I Townsend, January 8, 1827, Townsend Family Business, New York Public Library.

44. "Some Reminiscences," *Syracuse Times*, 1 July 1883, clipping in OHA Travelers Impressions file. See also Hand, *Forest to City*, 44; Smith, *Pioneer Times*, 289; Chase, *Syracuse and Its Environs*, 307. The Burnet House hotel in Cincinnati also issued local currency with its image on the bill. Boorstin, *National Experience*, 140.

45. Smith, *Pioneer Times*, 291.

46. Vandewater, *Tourist Guide*, 48. Of course, Vandewater probably got its write-up from a Syracuse correspondent. Steele, *Summer Journey in the West*, 36–37.

47. Cheney, "Reminiscences," 7. See also Hand, *Forest to a City*, 34–35.

48. Vandewater, *Tourist Guide*, 48.

49. [Jacques Milbert], "A French Naturalist Visits Rochester," ed. R. W. G. Vail, in *Centennial History of Rochester*, vol. 2., ed. Edward Foreman (Rochester: Rochester Public Library, 1931), 306–7.

50. O'Reilly, *Sketches of Rochester*, 376.

51. "Some Reminiscences," *Syracuse Times*, 1 July 1883, clipping in OHA Travelers Impressions file.

52. Letter from John Townsend to James McBride of Syracuse, 29 December 1838, Townsend Family Business Records, New York Public Library.

53. Being in the business of promoting Syracuse, the company actively supported individual enterprises, sometimes with financial assistance, as in the case of the Coffin Building, and other times with letters of recommendation. As explained in a letter written in support of John Wilkinson's request for credit from a Philadelphia firm, "We have no direct interest in this company, other than in the prosperity of the village." Letter from John and Isaiah Townsend to Messrs. A. and G. Rallston & Co., 26 January 1837, Albany, Townsend Family Business Records, New York Public Library.

54. Hand, *Forest to City*, 32–34; Evamaria Hardin, *Syracuse Landmarks: An AIA Guide to Downtown and Historic Neighborhoods* (Syracuse: Syracuse University Press, 1993), 30.

55. Vandewater, *Tourist Guide*, 48.

56. Longstreth, *Buildings of Main Street*, 24–29.

57. Hand, *Forest to City*, 100. See also Longstreth, *Buildings of Main Street*, 29–31.

58. "Improvements," April 1848, clipping in Raymond Scrapbook, 90. The architect Austin, who rented office space in the high-style Reynolds Arcade, also designed the new Monroe County courthouse in 1851.

59. Ibid.

60. "Notes and Incidents of Rochester; Letter No. 67," 19 April 1871, clipping in Scrantom Scrapbook, 1:46.

61. Vance, *Merchant's World;* Blumin, *Emergence of the Middle Class*, 78–80; Glenn Porter and Harold C. Livesay, *Merchants and Manufacturers: Studies in the Changing Structure of Nineteenth-Century Marketing* (Baltimore: Johns Hopkins Press, 1971), 13–36, 62–78.

62. In Rochester, where the canal was inserted into a growing city already a decade old, merchants had to adapt to the new artery. Buffalo Street proprietors altered their buildings into double-fronted businesses. Commercial buildings such as the prominent Talman Block were extended in the back to include an unornamented wing fronting Child's Basin. Bifurcated buildings were built by merchants, millers, and commission agents in other cities that also enjoyed two kinds of transportation advantage. Buildings along the water side of Troy's River Street, were subdivided into street-side retail-oriented businesses and river-side warehousing and shipping businesses. Still evident today, the facades at street level addressed common expectations of a pedestrian-oriented commercial district, with large glass windows and open fronts alternating with ornamental

jambs and lintels. The Hudson River side lacked any stylish gestures expected in a commercial district. Instead, its working river front displayed hoists, chutes, and bays for unloading the goods into the boats. Woodworth, *Reminiscences of Troy*, 24; Buel, *Troy for Fifty Years*, 13.

63. Not that the contents of this warehouse were necessarily secure; the building came to be known as the "Jerry Rescue Building" after citizens broke the jail open to free an arrested fugitive slave. Strong, *Early Landmarks*, 271–95.

64. "Reminiscences of Syracuse," *Syracuse Daily Standard*, 19 June 1867, clipping in OHA.

65. Public Meeting at Syracuse, 15 March 1833, recorded in Miscellaneous Canal Board Papers, New York State Archives.

66. "Syracuse Ordinances, 1851," sec. 14.

67. *Revised Charter of the City of Rochester, Passed April 11, 1844; To Which Are Added the City Ordinances* (Rochester: Erastus Shepard, 1844), sec. 7–9; *Act Incorporating the Village of Syracuse, Passed April 13, 1825, Also the Acts Amending the Same, Passed in the Years 1829, 1834, 1835, 1839 and 1841* (Syracuse, S. F. Smith, 1841), 25; *Charter of the City of Syracuse, Passed December 14, 1847; and Amendment Thereto Passed since that Time . . . and Also the Ordinances of the City of Syracuse* (Syracuse: Agan and Summer, 1851), 1849, sec. 40. See also *Acts of the Legislature of the State of New York, Relative to the City of Troy, and the Ordinances of the City of Troy* (Troy: Daily Steam Printing Office, 1870).

68. Upton, "The City as Material Culture," 56.

69. Rochester Ordinances, "An Ordinance Relating to Streets, Passed June 11, 1844," sec. 9. Generally, the city council devised a formula based on the width of the street. The wider the street, the larger the permissible encroachment: five feet on Rochester's main streets and three feet on the secondary streets. "Ordinances of the Village of Syracuse, in force June 19, 1841," 25. Troy similarly limited stoop and porch encroachments to six feet on the main streets. "A Law Relative to the Streets in the City of Troy, Passed February 8, 1838," sec. 7.

70. Rochester Ordinances, "An Ordinance Relating to Streets, Passed June 11, 1844," sec. 10; "Ordinances of the City of Syracuse, 1857," chap. 10, sec. 5. Troy similarly limited the projection of displays to three feet on its commercial streets. "A Law Relative to the Sidewalk on the East Side of River Street in the City of Troy, Passed May 20, 1853."

71. "Ordinances of the Village of Syracuse, in Force June 19, 1841," 23.

72. "Ordinances of the City of Syracuse, 1857," sec. 5.

73. "Ordinances of the Village of Syracuse, in Force June 19, 1841," 23. Rochester Ordinances, "An Ordinance Relating to Streets, Passed June 11, 1844," sec. 7.

74. Rochester Ordinances, "An Ordinance Relating to Streets, Passed June 11, 1844," sec. 9; "Ordinances of the City of Syracuse, 1857," chap. 10, sec. 5.

75. Rochester Ordinances, "An Ordinance Relating to Streets, Passed June 11, 1844," sec. 8. Troy passed a similar rule requiring that awnings be seven feet high, supported on square posts four inches wide and positioned at the outer edge of the sidewalk. "A Law Relative to the Streets in the City of Troy, Passed February 8, 1838," sec. 8.

76. "Ordinances of the Village of Syracuse, in Force June 19, 1841," 24; "Ordinances of the City of Syracuse, 1857," chap. 10, sec. 5.

77. [Hawthorne], "Sketches from Memory: Rochester," 407.

FIVE
Gentrifying the Sorted City

1. McIntosh, *History of Monroe County*, "Commercial and Mercantile Interests of Rochester," n.p.

2. Blackmar, *Manhattan for Rent*, 77–87; Johnson, *Shopkeeper's Millennium*; Bushman, *Refinement of America*; Blumin, *Emergence of the Middle Class*; Scobey, *Empire City*, 15–54.

3. Johnson, *Shopkeeper's Millennium*, 8.

4. Ibid., 27. See also Mahoney, *Provincial Lives*, 113–212.

5. Nancy A. Hewitt, *Women's Activism and Social Change: Rochester, New York, 1822–1872* (Ithaca: Cornell University Press, 1984), 50.

6. [Hawthorne], "Sketches from Memory: Rochester," 407.

7. Low, "Spatializing Culture," 113. In her analysis she draws particularly upon Michel Foucault. As Ian Hacking concludes, "the forms of knowledge and power since the nineteenth century have served the bourgeoisie above all others." "Archaeology of Foucault," 28, 30.

8. Low, "Spatializing Culture," 113. See also Carolyn J. Lawes, *Women and Reform in a New England Community, 1815–1860* (Lexington: University Press of Kentucky, 2000), 2.

9. Miles Ogborn, *Spaces of Modernity: London's Geographies, 1689–1780* (New York: Guilford Press, 1998), 109.

10. Low, "Spatializing Culture," 113. See also Michel de Certeau, *The Practice of Everyday Life* (Berkeley: University of California Press, 1984).

11. Low, "Spatializing Culture," 113–14; de Certeau, *Practice of Everyday Life*, xiv–xv. See also Susan Buck-Morss, *The Dialectics of Seeing: Walter Benjamin and the Arcades Project* (Cambridge: MIT Press, 1990), 35–40.

12. Scrantom Diary, 18 March 1856.

13. Cheney, "Reminiscences," 105.

14. Hand, *Forest to City*, 44; Smith, *Pioneer Times*, 287–92.

15. Charles Acton, "Tales: The President Stories; or Seven Nights at Welch's," *Literary Union* 1:2, 14 April 1849, 19.

16. Bushman, *Refinement of America*, 359.

17. Hand, *Forest to City*, 47.
18. Howison, *Sketches of Upper Canada*, 291–92. For a female perspective on upstate travel, see Caroline Gilman, *The Poetry of Travelling in the United States* (New York: S. Colman, 1838), 80.
19. Letter from Timothy C. Cheney to Ann Cheney, 23 October 1845, OHA Cheney Papers.
20. See, e.g., Fowler, *Journal of a Tour*, 72.
21. "Fifty Years After, Reminiscences of First and State Streets," *Troy Daily Press*, 4 November 1893; Weise, *Troy's One Hundred Years*, 121, 133.
22. "The Eagle—A Magnificent Hotel," 1848? clipping in Barton Scrapbook, 1:23; "George Darling," 1886? clipping in Barton Scrapbook, 1:107.
23. Strong, "Reminiscences for Early Rochester," 301.
24. "Ancient Landmarks," *Rochester Morning Herald*, 1887, clipping in Barton Scrapbook, 1:56. "The Clinton to Close," 1887, clipping in Barton Scrapbook, 1:98.
25. *Journal*, 2 February 1858, OHA Robbers Row File. See also "Syracuse as I Remember the City," *Post-Standard*, 30 March 1924, OHA Reminiscences file.
26. Augusta Rann Diary, 19 March 1861, Rann Daughter Diaries, OHA.
27. Scrantom Diary, 4 March 1851.
28. H. B. Skinner, "Correspondence—Buffalo, NY," *Christian Herald*, 26 October 1843, quoted in Sheriff, *Artificial River*, 119, 147.
29. William Oliver, *Eight Months in Illinois* (1843) (Great Americana Readex Microprint Corp., 1966), 60–62.
30. Peter Way, *Common Labor: Workers and the Digging of North American Canals, 1780–1860* (Baltimore: Johns Hopkins University Press, 1993), 164.
31. Of course, some groceries were ramshackle affairs, but not all. The prominent brothers James and John Leighton, born in Syracuse, went to work at a young age in the Stanton grocery store along the canal. By 1860 they had prospered sufficiently to have gone into business for themselves and constructed a large three-story building at the McBride Street lock crossing the canal. Builders designed these canal-focused enterprises to face the clientele. Because they serviced the canallers primarily, groceries such as the Leightons faced the canal side, with their merchandise or long row of display windows, and set up a more limited display on the street side. "Historic Houses of Syracuse," Scrapbook, 38, Onondaga County Public Library.
32. Sheriff, *Artificial River*, 139, quoting *Rochester Daily Advertiser*, 20 August 1830.
33. Quoted in Way, *Common Labor*, 163.
34. Ibid., 174.
35. Sheriff, *Artificial River*, 144, quoting *Laws of the State of New York, in Relation to the Erie and Champlain Canals, Together with the Annual Reports of the Canal Commissioners and Other Documents* (1825), 2:577–78.

36. "Syracuse Ordinances, 1841," 26.

37. Steele, *Summer Journey*, 47. See also Sheriff, *Artificial River*, 151–52.

38. "Petition of Inhabitants of Rochester, 11 March 1830," Canal Board Petitions, New York State Archives.

39. William Kenrick, *The Whole Duty of Woman, To Which Is Added Edwin and Angelina: A Tale* (Rochester: Everard Peck, 1819), 83, 68.

40. John F. Kasson, *Rudeness and Civility: Manners in Nineteenth-Century Urban America* (New York: Hill and Wang, 1990), 115.

41. Letter from Edwin Scrantom to Abelard Reynolds, Rochester, 27 January 1827, quoted in McKelvey, *Rochester: The Water-Power City*, 106.

42. Rochester Female Charitable Society, *Charter, Constitution, By-Laws*, 10–11, University of Rochester Rare Books and Special Collections.

43. Hewitt, *Women's Activism and Social Change*, 236.

44. Quotes from Charitable Society Papers, January 1838, quoted in Hewitt, *Women's Activism and Social Change*, 98.

45. Rochester Female Charitable Society, "Annual Report for the Year 1860," *Charter, Constitution, By-Laws*, 13–14; John C. Chumasero, *The Mysteries of Rochester* (Rochester: William H. Beach, 1845), 8.

46. "Syracuse Ordinances, 1846," sec. 22.

47. As the social historian Mary P. Ryan points out, "when either the cognitive or the institutional method of ordering public space failed, cities resorted to a third technique, the enactment of laws." *Women in Public: Between Banners and Ballots, 1825–1880* (Baltimore: Johns Hopkins University Press, 1990), 62.

48. "Rochester Ordinances, 1844," sec. 1; "Syracuse Ordinances, 1841," 24, 26; "Syracuse Ordinances, 1849," sec. 16; "Troy Ordinances, 1838," 475; "Troy Ordinances, 1846," 508; "Troy Ordinances, 1856," 571–72; "Troy Ordinances, 1858," 455.

49. Kasson, *Rudeness and Civility*, 116.

50. Kenrick, *Whole Duty of Woman*, 67.

51. American Lady, pseud., *True Politeness: A Hand-book for Ladies* (New York: George S. Appleton, 1848), 11.

52. Ibid., 6, 9.

53. Ibid., 12.

54. Hopkins, "Reminiscences of Miss Araminta D. Doolittle and the Rochester Female Academy," *Rochester Historical Society*, 133.

55. See also Ryan, *Women in Public*, 71–73.

56. "Hymn for Female Penitents," in *Trojan Sketchbook*, ed. Abba A. Goddard (Troy: Young and Hartt, 1846), 125.

57. John C. Chumasero, *Life in Rochester or Sketches from Life: Being Scenes of Misery, Vice, Shame, and Oppression, in the City of the Genesee, by a Resident Citizen* (Rochester: D. M. Dewey, 1848), 52, 91.

58. Phebe B. Davies, *Two Years and Three Months in the New-York State Lu-*

natic Asylum at Utica: Together with the Outlines of Twenty Years' Peregrinations in Syracuse (Syracuse: Evening Chronicle Book Bldg., 1855), 18.

59. "Syracuse Ordinances, 1857," chap. 3, sec. 2.
60. *True Politeness*, 14.
61. Rann Diary, 16 March 1861, 27 June 1861, 19 November 1861.
62. Ibid., 5 February 1861, 17 January 1861, 26 February 1861, 16 March 1861.
63. Ibid.
64. Alcesta F. Huntington Diary, 1 July 1850, 4 March 1852, 6 March 1852, 10 May 1866, 14 June 1866, 16 February 1866, 13 March 1866, Hooker Family Papers, Department of Rare Books and Special Collections, University of Rochester Libraries.
65. "An Awful Tragedy," *Rochester Democrat and American*, 21 December 1857.
66. Sarah and her brother Ira were convicted of the murder. Sarah served seven years in Sing Sing prison. Ira was hanged in the Monroe County jail. "Awful Tragedy"; Peck, *History of Rochester and Monroe County*, 1:172.
67. Huntington Diary, 13 March 1866. For a thoughtful analysis of the reshaping of privatized public space to sanction female participation in the city, see Abigail Van Slyck, "The Lady and the Library Loafer: Gender and Public Space in Victorian America," *Winterthur Portfolio* 31 (Winter 1996): 221–41.
68. Roberts, "Business Men of the Village," *Publications of the Onondaga Historical Association*, 31.
69. Letter from E. Cannon to Mary Warren, Troy, 5 August 183? Warren Family Papers, New York State Archives.
70. The male-oriented public spaces within the hotels forced some women to curtail their activities. After arriving at the Troy House, a female guest was put in ill humor by the immense and crowded dining hall and shortly retired to "a close and unhome-like bed-room." Gilman, *Poetry of Travelling*, 80.
71. Its furnishings were described as "almost equally handsome." "The Eagle—A Magnificent Hotel," 1848? clipping in Barton Scrapbook, 1:23. "George Darling," 1886? clipping in Barton Scrapbook, 1:107.
72. Leslie, "Western New York," 185.
73. Frances Kemble, *Journal Residence in America* (London: John Murray, 1835), 319; "Letter from Grace Greenwood," 1 August 1849, clipping in Raymond Scrapbook, 133; Whyte, "Diary," 42–43; Amelia M. Murray, *Letters from the United States, Cuba and Canada* (New York: G. P. Putnam, 1856), letter 12.
74. Low, "Spatializing Culture," 113; Ogborn, *Spaces of Modernity*, 109, 113–14.
75. Governor Clinton, *Speech of Gouvernor Clinton to the Legislature of the State of New-York* (Albany: Register Office, 1819), 11. He tolerated the idea of continuing the reservation lands in New York. Clinton did not advocate forced relo-

cation but hoped the Indians would move along on their own volition. Hamer, *New Towns in the New World*, 217. In other accounts whites noted that the "good indians" were to be found in the reservations. E. S. Abdy, *Journal of a Residence and Tour in the United States of North America, from April, 1833, to October, 1834* (1835; rpt., New York: Negro Universities Press, 1969), 329. Such attitudes, of course, assumed that Native Americans wanted to be in town. Handsome Lake's attempts to consolidate the smaller reservations into the larger ones was motivated in part, as Wallace has explained, by a desire to avoid dissolute white settlements. Wallace, *Death and Rebirth of the Seneca*, 280, 208–11.

76. "Scrantom Letter No. 27," Scrantom Scrapbook, 1:20.

77. Scrantom's observations were typical. See, e.g., Combe, *Notes on the United States*, 2:324–26.

78. [Milbert], "French Naturalist Visits Rochester," 307–8; Milbert, *Picturesque Itinerary*, 129.

79. Bernard W. Sheehan, *Savagism and Civility: Indians and Englishmen in Colonial Virginia* (Cambridge: Cambridge University Press, 1980), ix–x; Hamer, *New Towns in the New World*, 216–17.

80. James Hall, *Sketches of History, Life, and Manners in the West* (Philadelphia: Harrison Hall, 1835), 1:27–133, quoted in Pearce, *Savagism and Civilization*, 72.

81. Sheehan, *Savagism and Civility*, x. See also Ogborn, *Spaces of Modernity*, 75, for the way a "racialised discourse of savagery" framed concerns about disorder in the built environment. See also Hamer, *New Towns in the New World*, 216.

82. O'Reilly, *Sketches of Rochester*, 275–77; Wallace, *Death and Rebirth of the Seneca*, 50–54, 245, 293, 335–36; "Letter from Syracuse," *Syracuse Standard*, 7 July 1857, clipping from OHA Travelers Impressions file; Elisabeth Tooker, "Iroquois since 1820," in *Handbook of the North American Indians*, vol. 15: *Northeast*, ed. Bruce G. Trigger (Washington, D.C.: Smithsonian Institution, 1978), 457.

83. Wallace, *Death and Rebirth of the Seneca*, 208.

84. O'Reilly, *Sketches of Rochester*, 277. The white dog ritual was also performed in Onondaga County. Recalling her childhood, Elizabeth Steven Huntington, born in 1806 near Onondaga Valley, remembered "the burning of the white dog" festivals, noting that "the white settlers were incited to attend. Those whites who were not present at these celebrations were not looked on with favor by the red men." "At Onondaga Valley—Mrs. Huntington Recalls Reminiscences of the Early Settlement There," *Syracuse Herald*, 1 September 1886, OHA Reminiscences file. The ritual continued on the Iroquois reservations at least until the 1870s. "Feast of the White Dog," *Syracuse Herald*, 1914, in "Stories of Early Syracuse and Onondaga County, Syracuse Herald, 1914–1916," 4, Onondaga County Public Library.

85. O'Reilly, *Sketches of Rochester*, 275–77; Barber and Howe, *Historical Col-*

lections (1841), 268; John W. Barber and Henry Howe, *Historical Collections of the State of New York: Being a General Collection of the Most Interesting Facts, Biographical Sketches, Varied Descriptions, &C. Relating to the Past and Present: With Geographical Descriptions of the Counties, Cities, and Principal Villages throughout the State* (New York: Clark, Austin, 1851), 371, 374.

86. Barber and Howe, *Historical Collections*, 389.
87. Milbert, *Picturesque Itinerary*, 129.
88. Clinton, *Speech of Gouvernor Clinton . . . 1819*, 11.
89. Hamer, *New Towns in the New World*, 217.
90. Sheehan, *Savagism and Civility*, 179; Roy Harvey Pearce, *Savagism and Civilization: A Study of the Indian and the American Mind* (Berkeley: University of California Press, 1988), 59–60. For the contrast of the lack of drunkenness on the reservations, see Wallace, *Death and Rebirth of the Seneca*, 303–10.
91. "Scrantom Letter No. 27," Scrantom Scrapbook, 1:20.
92. Ibid.
93. Hawthorne, for instance, noted a drunken party of Scotch and Irish military recruits. Weber, "Hawthorne's American Travel Sketches" (1832), 46–47; Alfred Weber, Beth L. Lueck, and Dennis Berthold, *Hawthorne's American Travel Sketches* (Hanover, N.H.: University Press of New England, 1989), 46–47. See also Capt. Francis Dana Watchbook, 1836–1838, Rochester Public Library.
94. "Scrantom Letter No. 27," Scrantom Scrapbook, 1:20.
95. Ibid.
96. Johnson, *Shopkeeper's Millennium*, 55–61.
97. "Ordinances of the City of Syracuse, as Revised and Adopted July 2, 1849," sec. 6.
98. Bremer, *Homes of the New World*, 1:580–81.
99. "Reminiscences of Mrs. Elisha Sibley, 1891? Barton Scrapbook, 1:35.
100. [Thomas K. Wharton], "From England to Ohio, 1830–1832: The Journal of Thomas K. Wharton," ed. James H. Rodabaugh, *Ohio Historical Quarterly* 65:1 (January 1956): 22.
101. Leslie, "Western New York," 185.
102. Jenks, *Notes on a Tour through the Western Part of the State of New York*, 26. See also Alvin Fisher, "Remnant of the Tribe Leaving the Hunting Ground of Their Fathers," in Delaware Art Museum and illustrated in Patricia A. Junker, *The Course of Empire: The Erie Canal and the New York Landscape, 1825–1875 : June 16–August 12, 1984, Memorial Art Gallery of the University of Rochester, Rochester, New York: Exhibition and Catalogue* (Rochester: Gallery, 1984).
103. Addendum to Cheney, "Reminiscences," 184. Pay day likely referred to the government annuities paid according to the various treaties in which they sold their lands. For example, in 1788 the Onondaga ceded all of their lands to New York except for a 100–square mile tract; in 1793 they sold about three-fourths of this reservation; in 1795 they sold their rights to Onondaga Lake; in

1817 they sold a 4,320-acre tract east of the reservation; in 1822 they sold 800 more acres. In payment the Onondaga received $33,380 in cash, $1,000 in clothing, and an annuity of $2,430, and 150 bushels of salt. Harold Blau et al. "Onondaga," in *Handbook of the North American Indians*, vol. 15: *Northeast*, 496. According to the U.S. Treaty of Canandaigua, there were additional annuities payable to all Iroquois. Wallace, *Death and Rebirth of the Seneca*, 218. See also Susan E. Gray, *The Yankee West: Community Life on the Michigan Frontier* (Chapel Hill: University of North Carolina Press, 1996), 67–71.

104. Buckingham, *America*, 2:255.

105. "Talk of Old Times, 1892," Barton Scrapbook, 1:52; Buckingham, *America*, 2:255. In Troy Mohawks also played on their appeal as exotics. "Trojan Reminiscences," *Troy Daily Press*, July 1894.

106. Taylor, *William Cooper's Town*, 33–34; William T. Hagan, "Justifying Dispossession of the Indian: The Land Utilization Argument," in *American Indian Environments: Ecological Issues in Native American History*, ed. Christopher Vecsey and Robert W. Venables (Syracuse: Syracuse University Press, 1980), 65–80; Robert W. Venables, "Iroquois Environments and 'We the People of the United States,'" in *American Indian Environments: Ecological Issues in Native American History*, ed. Christopher Vecsey and Robert W. Venables (Syracuse: Syracuse University Press, 1980), 81–127; Pearce, *Savagism and Civilization*, 65–68, 101–2, 240.

107. Myers, *Sketches on a Tour*, 113.

108. Scrantom Scrapbook, 1:1; "Local Affairs," *Democrat and American*, 16 June 1862.

109. Chumasero, *Life in Rochester*, 48.

110. *Directory of the City of Rochester for 1845–6* (Rochester: Canfield and Warren, 1845), 18; Vance, *Merchants World*, 29–32, 45–46, 105.

111. Chumasero, *Mysteries of Rochester*, 4.

112. Chumasero, *Life in Rochester*, 84.

113. *Rochester City Directory 1844*, 6; "Was Given to Rhyming," *Syracuse Herald*, 6 June 1894; "Principal Block in 1847 Merely Memories Now," *Syracuse Journal*, 14 April 1925.

114. "Syracuse, It's Rapid Growth," *Syracuse Journal*, 29 February 1872.

115. Hand, *Forest to City*, 28–30.

116. Ibid., 29. "To carry off the palm" was a common nineteenth-century phrase meaning to be the best, the winner, or what we might today colloquially call being "first fiddle."

117. Domosh, *Invented Cities*, 35–64; Susan Porter Benson, *Counter Cultures: Saleswomen, Managers, and Customers in American Department Stores, 1890–1940* (Urbana: University of Illinois Press, 1986), 7; Shirley T. Wajda, "Social Currency: A Domestic History of the Portrait Photograph in the United States, 1839–1889" (Ph.D. diss., University of Pennsylvania, 1992), 334–452; Hunting-

ton Diary, 6 March 1852; Rann Diary, 2 April 1861, 16 July 1861, 24 July 1861; Foreman, *Centennial History of Rochester*, 2: facing 58.

118. "To Rent," *Rochester Daily Democrat*, 19 July 1850.

119. Blumin, *Emergence of the Middle Class*, 98–107.

120. Leslie, "Western New York," 185.

121. Ryan, *Women in Public*, 76–77.

122. Quoted in Hunt, "Late Hours of Business," *Wealth and Worth*, 117.

123. Hunt, "Investment," 127.

124. "Jerome Reminiscences," 6 March 1891, clipping in Barton Scrapbook, 1:116.

125. Benson, *Counter Cultures*, 5, 7.

126. Hunt, "Employment of Ladies as Clerks in Stores," 499. See also Benson, *Counter Cultures*, 23.

127. Gunther Barth, "Demopiety: Speculations on Urban Beauty, Western Scenery, and the Discovery of the American Cityscape" *Pacific Historical Review* 52:3 (August 1983): 255; *City People: The Rise of Modern City Culture in Nineteenth-Century America* (New York: Oxford University Press, 1980), 110–47; Alan Trachtenberg, *The Incorporation of America: Culture and Society in the Gilded Age* (New York: Hill and Wang, 1982), 130–35; Benson, *Counter Cultures*, 12–30.

128. Floor plan of Burke, Fitzsimmon, Hone & Co., *Rochester City Directory, 1861*, 587–88. By the end of the decade business had picked up enough to warrant sixty-five salesmen. McKelvey, *Rochester: The Flower City*, 163; "To Rent," *Rochester Daily Democrat*, 19 July 1850. The upper floor housed the Odd Fellows Hall. "Emporium Buildings," 1847? clipping in Raymond Scrapbook, 90.

129. Hand, *From a Forest to a City*, 29–30.

130. Letter from Lois Freeman to Susan Crary, Syracuse, 23 November 1840, OHA.

131. Chumasero, *Life in Rochester*, 44.

132. Edward Hazen, *The Panorama of Professions and Trades; or Every Man's Book* (Philadelphia: Uriah Hunt, 1837).

133. Blumin, *Emergence of the Middle Class*, 76–78.

134. Chumasero, *Life in Rochester*, 12.

135. Strong, "Reminiscences of Early Rochester," 298.

136. Hunt, "Mercantile Education," 197. The exchange from unstructured playground to ordered time and space played out in personal as well as architectural fashion. Hunt continued: "His easy dress must undergo a change also. The round jacket must give place to a premature long-tailed coat; the easy shoe the high heeled boot; the open shirt collar to the starched cravat."

137. "News and Incidents of Rochester, No. 19," 1:15.

138. *Boyd's Rochester Directory with a Business Directory* (Rochester: Andrew Boyd, 1863), 42.

139. Hazen, *Panorama of Professions and Trades*, 61, 42, 30, 189, 252.

140. Clayton Mau, *The Development of Central and Western New York from the Arrival of the White Man to the Eve of the Civil War as Portrayed Chronologically in Contemporary Accounts* (Dansville, N.Y.: Own Publishing, 1958), 315, quoting *Livingston Register*, 23 October 1838.

141. Chumasero, *Life in Rochester*, 68.

142. *Boyd's Rochester Directory with a Business Directory* (Rochester: Boyd, 1861), 9.

143. "The Littles Murder," *Rochester Democrat and American*, 22 December 1857.

144. By 1820 Rochester's total population of 1,502 included 18 free blacks and 9 slaves; by 1860 the number had risen to 410 free blacks in a city of 48,204 residents. In 1820 the approximately 600 residents of Salina township, which included Syracuse, had 19 African Americans, including four slaves; by 1840 Syracuse's approximately 6,500 residents included 234 free blacks and no slaves. McKelvey, *Rochester: The Flower City*, 3; *Rochester: The Water-Power City*, 69; Eugene Du Bois, *The City of Frederick Douglass: Rochester's African American People and Places* (Rochester: Landmark Society of Western New York, 1994), 6; Barbara Shelkin Davis, *A History of the Black Community of Syracuse* (Syracuse: Onondaga Community College, 1980), 5. The Englishman James Silk Buckingham found the absence of blacks and Indians striking: "among the minor peculiarities of Rochester, we remarked that there were fewer people of colour seen in the streets than in any town we had visited." *America*, 2:215.

145. Howard W. Coles, ed., *The Cradle of Freedom: A History of the Negro in Rochester, Western New York, and Canada* (Rochester: Oxford University Press, 1941), 1:78–79.

146. *Daily American Directory of the City of Rochester for 1848–50* (Rochester: Jerome and Brother, 1849); Du Bois, *City of Frederick Douglass*, 10–15.

147. Frederick Douglass, *The Life and Times of Frederick Douglass: His Early Life as a Slave, His Escape from Bondage, and His Complete History* (1892) (New York: Collier Books, 1962), 264.

148. "Progress of Justice and Equality," *North Star*, 23 June 1848.

149. Ibid.; Coles, *Cradle of Freedom*, 1:133; Philip S. Foner, *The Life and Writings of Frederick Douglass, Early Years, 1817–1849* (New York: International Publishers, 1950), 85. Ironically, in 1851 the Irving House was the site for a reception for the Hungarian freedom fighter Louis Kossuth, who ceremonially received the delegation of African Americans, who praised him for upholding "the principle that a man has the right to the full exercise of his faculties and powers in the land which gave him birth." "Reception of Colored Persons," *Frederick Douglass' Paper*, 18 December 1851.

150. Lindley E. Gould, "Diary and Account Book of Lindley E. Gould, 1817–1853," 14 May 1842, Rochester Public Library.

151. H. Perry Smith, *The Royal Road to Wealth: An Illustrated History of the Successful Business Houses of Syracuse* (Syracuse: Vanarsdale, 1873), 8–14; Marjorie H. Thorpe, "The Story of Renwick Castle, 1852–1932," 2–12, clipping in Onondaga County Public Library; "Renwick Castle," in "Stories of Early Syracuse and Onondaga County, from the *Syracuse Herald*, 1914–1916," 74, Onondaga County Public Library.

152. Smith, *Royal Road to Wealth*, 9–12.

153. Queried one newspaper article "Do you remember the 'stovepipe' hat gentry of those days.... It was the badge of gentility, that silk hat." "Syracuse in the Sixties," *Syracuse Post-Standard* clipping, OHA Reminiscences file.

154. Smith, *Royal Road to Wealth*, 12.

155. Ibid., 10–11.

156. Ibid., 11.

157. Store tokens were privately minted during the coin shortage during the Civil War. In addition to making spare change, they also were convenient forms of advertisement. The Yates token had on its obverse an eagle and the Salina Street address of the store. Yates store token in author's private collection.

SIX
The Reynolds Arcade and Athenaeum

1. "From the *Titusville Herald*," in *Address and Sermon on the Death of William Abelard Reynolds by M. B. Anderson, LL.D., and Rev. D. K. Bartlett with Other Memorial Papers* (Rochester: E. R. Andrews, 1872), 65–66.

2. In his collection of "maxims, morals, and miscellanies" for businessmen, the publisher Freeman Hunt reiterated location as the key consideration for a shopkeeper: "it is essential to the success of a retail tradesman, to establish himself in some leading thoroughfare." "Choice of a Store," *Worth and Wealth*, 47.

3. Political allegiances also lost Reynolds the postmaster position the year after the Arcade opened. The post office remained in the Arcade nonetheless. Johnson, *Shopkeeper's Millennium*, 28–30; Amy Hanmer-Croughton, "Historic Reynolds Arcade," *Rochester Historical Society Publication Fund Series* 8 (1936): 97–108; Reynolds, "Autobiography," *Rochester Post-Express*, 23 September 1884; *Illustrated Guide to Reynolds Arcade: With History of the Arcade from 1828, Plans of Each Floor, Number of Each Office, Names and Business of Each Occupant* (Rochester: George Frauenberger, 1880).

4. Johnson, *Shopkeeper's Millennium*; Jenny Marsh Parker, *Rochester: A Story Historical* (Rochester: Scrantom, Wetmore and Co., 1884), 283; Vandewater, *Tourist Guide*, 54.

5. Vandewater, *Tourist Guide*, 54.

6. Robert Alexander, "The Arcade in Providence," *Journal of the Society of Architectural Historians* 12:3 (October 1953): 13–16. It is unclear how much knowl-

edge Reynolds had of specific arcade prototypes. The Providence Arcade was illustrated a few years after the construction of the Reynolds Arcade in the nationally distributed *American Magazine of Useful Knowledge* (1834): 303. In 1979 the arcade historian Johann Geist tentatively called the Rochester Arcade a copy of London's 1819 Burlington Arcade. Upon examination, however, neither the extraordinary length of the Burlington Arcade nor its arched street or shop entrances were evident in Reynolds Arcade. Johan Friedrich Geist, *Arcades: The History of a Building Type*, trans. Jane O. Newman and John H. Smith (Cambridge: MIT Press, 1983), 12–16, 543, f. 745; M. R. G. Conzen, "The Plan Analysis of an English City Centre," in *Proceedings of the I.G.U. Symposium in Urban Geography, Lund*, ed. Kurt Norberg (Lund, Sweden: CWK Gleerup, 1962), 388.

7. *Illustrated Guide to Reynolds Arcade*, 3–10; Vandewater, *Tourist Guide*, 54.

8. "Ancient Landmarks," *Rochester Morning Herald*, 1887, clipping in Barton Scrapbook, 1:56.

9. Letter from William Reynolds to Abelard Reynolds, 2 September 1846, in Blake McKelvey, "Letters Postmarked Rochester," *Rochester Historical Society Publications*, 21 (1943): 86–87.

10. Letter from William Reynolds to Abelard Reynolds, 27 September 1846, in "Letters Postmarked Rochester," 88–89.

11. Ibid., 89; "Opinions of Rochester," 1847 (?), newspaper clipping in Raymond Scrapbook, 56.

12. Abelard Reynolds (1785–1879) moved to Rochester in 1813. William Reynolds (1810–72) assumed control of his father's Arcade from 1845 until his death. Mortimer Reynolds, William's brother, then ran the business until 1892.

13. Letter from William Reynolds to Abelard Reynolds, 5 November 1846, in "Letters Postmarked Rochester," 90.

14. Letter from William Reynolds to Abelard Reynolds, 2 September 1846, in ibid., 87–88.

15. Ibid.

16. By 1839 the library held approximately four thousand books. McKelvey, *Rochester: The Water-Power City*, 194, 274, 345.

17. Letter from William Abelard to Robert Denniston, 10 June 1848, in "Letters Postmarked Rochester," 94. Reynolds was selling his telegraph stock, refinancing his debt payments, and trying to convince his mortgage holder, William B. Astor of New York City, to extend the terms of his payments.

18. Letter from William Reynolds to William Astor, 20 January 1849, in ibid., 95.

19. "City Improvements," 1848, clipping in Raymond Scrapbook, 90.

20. The building was remodeled again in 1879 and torn down during the 1920s to be replaced by a new Arcade building.

21. "City Improvements," 90.

22. A fuller description of the Corinthian Hall and its events in found in Elwood, *Public Amusements*, 45–52.

23. Scrantom Scrapbook, vol. 1: "Notes and Incidents of Rochester. Letter No. 67," 19 April 1871, 46. This idea had already been raised when the Athenaeum was under construction. Raymond Scrapbook, "Improvements," 90.

24. Raymond Scrapbook, "Improvements," 90.

25. James Mathies, *Rochester, a Satire: And Other Miscellaneous Poems* (Rochester: n.p., 1830), 15.

26. Glynn, "The Rochester Arcade Quick Step," University of Rochester, Rare Books and Special Collections; Josiah W. Bissell, "Reminiscences of Early Rochester," *Rochester Historical Society Publication Fund Series* 6 (1927): 305.

27. Bremer, *Homes of the New World*, 1:586. See also Patrick Sheriff, *A Tour through North America; Together with a Comprehensive View of the Canadas and United States. As Adapted for Agricultural Emigration* (Edinburgh: Oliver and Boyd, 1835), 86.

28. Cleveland, "Journal of a Tour, 1831," 358.

29. McKelvey, *Rochester the Flower City*, 23.

30. Parker, *Story Historical*, 155.

31. Letter from William Reynolds to the Sec. Fireman's Insurance Co., 8 June 1850, in "Letters Postmarked Rochester," 97.

32. "Presbyterian Church in Rochester," *American Magazine of Useful Knowledge* (1836), 323.

33. Pred, *Urban Growth and City-Systems*, 142–65; Vance, *Merchant's World*, 156–57; Cronon, *Nature's Metropolis*, 293–94; *Rochester Directory 1863*, 62–64.

34. "The Arcade Hall," 1851, clipping in Raymond Scrapbook, 156. See also Buck-Morss, *Dialectics of Seeing*, 35–40.

35. "From the *New York Evangelist*," in *Address and Sermon on the Death of William Abelard Reynolds*, 65.

36. My thanks to Shirley Wajda for this idea. The name Corinthian Hall was popular in other mercantile cities as well, including Syracuse.

37. McKelvey, *Rochester: The Flower City*, 34, 361.

38. *Boyd's Rochester Directory 1863–64*, 62–64.

39. *Rochester Daily Union Annual City Directory* (Rochester: Daily Union, 1859), n.p.; *Boyd's Rochester Directory, 1863–64*, 62–64.

40. *History of Monroe County*, 110. For ladies in the reading rooms, see Huntington Diary, 16 February 1866, 13 March 1866; *True Politeness*, 64.

41. Frederick A. Whittlesey, "Remembrance and Prophesy: Tales of a Grandfather," *Centennial History of Rochester New York*, 2:28. A clipping in the Raymond scrapbook also heralded the social as well as architectural improvement of the alley. "This will be a completion of the renovation of old 'Bugle Alley,' whose haunts and memories are classical in the minds of many who used to keep vigil

there for the safety of the city, and from their love for their corps of ardent firemen." Raymond Scrapbook, "Improvements," 90.

42. Scrantom Scrapbook, "Notes and Incidents of Rochester," No. 27, 1:20; Scrantom Scrapbook, "Notes and Incidents of Rochester in the Old Time and the New, by an Old Citizen," no 12, 1:11.

43. *Records of the Doings of the Trustees of the Village of Rochesterville*, 7 May 1818.

44. Capt. Francis Hall Watchbook, n.p., Rochester Public Library.

45. "Matrimony!" *Frederick Douglass' Paper*, 9 November 1855. Jenny Lind's sensational tour through Rochester in 1851 prompted unheard-of ticket prices, including a show sold at auction, after which Lind donated over twenty-five hundred dollars to the city charities, including eight hundred dollars to the Female Charitable Society. George M. Elwood, *Some Earlier Public Amusements of Rochester* (Rochester: Democrat and Chronicle, 1894), 51.

46. "Arcade Hall," 1851, clipping in Raymond Scrapbook, 156. The office of watchman was specifically created to combat the crimes of the night. "[Watchmen] are empowered to patrol the streets, roads, alleys, yards & every place other in said Village at any & all times between the setting of the sun & rising of the same; and to examine any & all houses, out-houses, stables, yards, buildings and tenements whatsoever in said Village—and to arrest, detain & examine any suspicious person found in said Village at any time between sunset & sunrise." *Records of the Doings of the Trustees of the Village of Rochesterville*, 7 May 1818.

47. "From the *Titusville Herald*," in *Address and Sermon on the Death of William Abelard Reynolds*, 66.

48. "Remarks of President Anderson," in *Address and Sermon on the Death of William Abelard Reynolds*, 11; "Sermon of Rev. Bartlett," in *Address and Sermon on the Death of William Abelard Reynolds*, 26.

49. "Meeting of Arcade Occupants," in *Address and Sermon on the Death of William Abelard Reynolds*, 41.

50. "From the *Rochester Democrat and Chronicle*," in *Address and Sermon on the Death of William Abelard Reynolds*, 44.

51. "Meeting of the Common Council," in *Address and Sermon on the Death of William Abelard Reynolds*, 44.

52. "From the *Rochester Union and Advertiser*," in *Address and Sermon on the Death of William Abelard Reynolds*, 59.

53. "Sermon of Rev. Bartlett, in *Address and Sermon on the Death of William Abelard Reynolds*, 32.

54. Ibid., 23.

55. "Remarks of President Anderson," in *Address and Sermon on the Death of William Abelard Reynolds*, 11.

56. "From the *Rochester Democrat and Chronicle*," in *Address and Sermon on the Death of William Abelard Reynolds*, 55, 57.

NOTES TO PAGES 138–143 195

57. Ibid., 57; "From the *Rochester Union and Advertiser,*" in *Address and Sermon on the Death of William Abelard Reynolds,* 44.
58. "Sermon of Rev. Bartlett," in *Address and Sermon on the Death of William Abelard Reynolds,* 23.
59. "Meeting of Arcade Occupants," in *Address and Sermon on the Death of William Abelard Reynolds,* 42; "Athenaeum and Mechanics' Association," in *Address and Sermon on the Death of William Abelard Reynolds,* 52.
60. "Meeting of Arcade Occupants," in *Address and Sermon on the Death of William Abelard Reynolds,* 41–43.
61. "Funeral Trappings and Suits of Woe," in *Address and Sermon on the Death of William Abelard Reynolds,* 70–71; "Meeting of Arcade Occupants," in *Address and Sermon on the Death of William Abelard Reynolds,* 41–43.

SEVEN
Transportation and the Changing Streetscape

1. Sheriff, *Artificial River,* 64, quoting Everard Peck to Samuel Porter, 22 April 1828.
2. The weighlock brought in $60,000 a year, forwarding lines brought in about $320,000, and other canal revenues were calculated at $192.000. "Presbyterian Church in Rochester," *American Magazine of Useful Knowledge* 3 (1836): 323.
3. Sheriff, *Artificial River,* 95.
4. O'Reilly, *Sketches of Rochester,* 233–34.
5. Clapp, "Travel, Trade, and Transportation," 180–81.
6. The New York State Archives Canal Board records hold numerous petitions on this topic. See, e.g., "Robert Furman claim for damages, 1838; James O. Bennett, 1838; McBride claims for damages, 1843; Tavvey, 1843; and Davis, 1843.
7. Petition of Jonathan Swift, 1841, Canal Petitions, New York State Archives.
8. Hervey Ely Petition, 20 August 1840, Canal Petitions, New York State Archives.
9. See, e.g., Robert Furman of Syracuse claim for damages, 1838, New York State Archives Canal Board.
10. "Letter No. 19; News and Incidents," Scrantom Scrapbook, 1:15–17.
11. Peck, *Landmarks of Monroe County,* 140–42; Whittlesey, "Tales of a Grandfather," 27.
12. "Letter No. 19; Notes and Incidents," Scrantom Scrapbook, 1:15.
13. Strong, "Reminiscences of Early Rochester," 300.
14. Peck, *Landmarks of Monroe County,* 170; *Daily American Directory of the City of Rochester for 1849–50* (Rochester: Jerome and Brother, 1849), 256.
15. [Scrantom], "And This Was Rochester! *Rochester History,* 20.

16. Ibid.

17. *Syracuse Daily Standard*, 7 April 1850, quoted in Craig Williams, "The Syracuse Weighlock Building" (Master's thesis, SUNY-Oneonta, 1983), 61.

18. Roberta Balstad Miller, *City and Hinterland: A Case Study of Urban Growth and Regional Development* (Westport, Conn.: Greenwood Press, 1979), 86, quoting Stevens, *Beginnings of the New York Central*, 149.

19. Vance, *North American Railroad*, 84–88.

20. "Syracuse, New York," *Ballou's Pictorial Drawing Room Companion* 11:5, 2 August 1856, 72–73.

21. Residents did complain to the canal board about stagnant canal waters even during the first decade of the canal, although the complaints were typically couched as part of a request to draw off some of the water for the petitioner's own personal use. "Petition of David H. Carter, 1828," Canal Records, New York State Archives.

22. Chase, *Syracuse and Its Environs*, 1:395–96.

23. Carroll L. V. Meeks, *The Railroad Station: An Architectural History* (New Haven: Yale University Press, 1956), 30, 50.

24. Van Slyck, "Lady and the Library Loafer," 237–38.

25. "The Citizen's Coffee House and Ice Cream Saloon," 24 May 1843, in Tredwell Scrapbook, 457, OHA; Townsend Family Business Records, box 32, Syracuse Company Miscellaneous Papers, 1827–1845, New York Public Library.

26. This would have been similar to what had been done in 1835 in Troy, where the tracks ran down the middle of River Street, giving easy access to shipping and commercial businesses.

27. Cheney, "Reminiscences"; Clark, *Onondaga*, 2:92; Hand, *Forest to a City*, 71; "Historical Papers, No. 17," OHA *Herald* Roads Collections; Smith, *Pioneer Times in Onondaga County*, 245; Conzen, "Plan Analysis," 392.

28. The federal government, however, considered Wilkinson's position as Syracuse's longtime contract as the postmaster a conflict of interest because the railroad carried the mail. He resigned as postmaster rather than give up the more profitable railroad. Clayton, *History of Onondaga County*, f. 150.

29. Hand, *Forest to a City*, 74.

30. "Our City: Impressions of Syracuse," *Syracuse Journal*, 8 June 1869, OHA Travelers Impressions file.

31. "The Central City," *Syracuse Journal*, 23 April 1870, OHA Travelers Impressions file.

32. Ibid.

33. "Letter from Syracuse," *Syracuse Standard*, 25 July 1857, OHA Travelers Impression file.

34. Chudacoff and Smith, *Evolution of American Urban Society*, 62–63. For an analysis of downtown railroads in Philadelphia, see also, Jeffrey P. Roberts, "Railroads and the Downtown: Philadelphia, 1830–1900," in *The Divided Metropolis: So-*

NOTES TO PAGES 149–150   197

cial and Spatial Dimensions of Philadelphia, 1800–1975, ed. William W. Cutler III and Howard Gillette Jr. (Westport, Conn.: Greenwood Press, 1980), 27–55.

35. Such a trend required economic maturity, predictability, and access to markets that had not been secured in the earlier days. As the towns grew, the local market demand strengthened. As industrialization increased, more households switched from home production to store-bought goods. As transportation continued to improve, merchants were able to stock their stores reliably with more and varied goods. As a result, merchants were increasingly able to specialize in retailing alone and faced new opportunities for the economic and physical expansion of their business. Vance, *Continuing City*, 406–7; *Merchant's World*, 23–24, 29–32, 45–46, 105.

36. "Report of the Committee on Public Lands, on the Petition of the President and Trustees of the Village of Syracuse," 12 March 1842, Records of the State of New York, no. 106, 5.

37. "Report of the State Engineer and Surveyor," 26 February 1849, Records of the State of New York, no. 130, 4.

38. The Syracuse Company, however, was not yet prepared to give up the mill or to bring those lots into the marketplace. John Townsend of the Syracuse Company defended keeping the pond, noting that he had as much, if not more, at stake in Syracuse than many of the residents: "I do consider that I have a much larger amount of property to be employed lying in the immediate vicinity of the pond than any other." He accepted, however, that it was "my duty to use my best exertion to improve the condition of this pond." Townsend's plan was to try to keep the pond at a continuous level rather than permitting draw-downs that exposed the swamp. By repairing the edges of the pond and stopping the mills during the dry months, he was convinced, "all the difficulties about the mill pond should never again be complained of." Letter from John Townsend to (Jonas?) Wescot, 11 April 1848, Townsend Family Business Records, New York Public Library.

39. Report filed with the *Map of Onondaga Creek and Adjoining Lands*, B. Green, 1849; Canal Records, New York State Archives; "Syracuse 1849 Ordinances," chap. 12; Bruce, *Memorial History of Syracuse*, 143.

40. Henry Schramm, *Central New York: A Pictorial History* (Norfolk: Donning, 1987), 37.

41. Letter from Allan Munroe to John Townsend, 24 May 1853, Townsend Family Business Records, New York Public Library.

42. The first arsenal was completed in 1858, burned in 1871, and was rebuilt again in 1872–74 by Horatio Nelson White. In 1903 it was replaced by the armory, which still stands today. Horning, *Horatio Nelson White*, 19–20.

43. "Our City," *Syracuse Journal*, 8 June 1869, OHA Travelers Impressions file.

44. Ultimately, the legacy of this sequestering of production away from the public sight in the commercial district found its greatest expression in the separation of manufacturer's factories and their sales offices. During the 1860s white-

collar and blue-collar employees of the same business began to be sorted out into their separate districts. For example, by 1862 in Syracuse, Ellis, Wicks & Company's leather warehouses were located at 33 West Water Street, beyond Clinton Square, but its sales offices were at 17 East Water Street by Hanover Square. *Syracuse Daily Journal City Directory for 1862–63* (Syracuse: Truair, Smith and Miles, 1862), 17.

45. This pattern in southern cities has been noted by Tolbert, "Constructing Townscapes," 125, 134.

46. Hunt, *Worth and Wealth*, 47.

47. W. W. Clayton, *History of Onondaga County, New York* (Syracuse: D. Mason, 1878), 227, 246; Strong, "Merchants in Exchange Street," 298–99; "Syracuse, Its Rapid Growth," *Syracuse Journal*, 29 February 1872; "Was Given to Rhyming, Seth William's Way of Advertising," 6 June 1894, newspaper clipping, OHA Reminiscences File; Clayton, *History of Onondaga County*, 246; Schramm and Roseboom, *Syracuse*, 46; "In Forty-Eight," *Syracuse Herald*, 14 March 1897, OHA Reminiscences file; "Syracuse: The Rapid Growth," *Syracuse Journal*, 29 February 1872.

48. Lewis Bradley, the artist of the Syracuse view, had exhibited at the National Academy of Design in New York City. The peculiarities of perspective, scale, and confusingly overlaid buildings in the Syracuse view may explain his short career as a city painter. Only three views are attributed to him: his hometown of Utica; Syracuse; and its rival, Oswego. John W. Hill, the artist of the Rochester view, worked for the New York State Geological Survey before he began drawing cities. The Rochester lithograph is one of twenty-seven known views from thirteen states by Hill, who later struck out on his own. John W. Reps, *Views and Viewmakers of Urban America: Lithographs of Towns and Cities in the United States and Canada, Notes on the Artists and Publishers, and a Union Catalog of their Work, 1825–1925* (Columbia: University of Missouri Press, 1984), 183–84, 206–8.

49. French, *Historical and Statistical Gazetteer*, 109–10, 406, 490.

50. Over time lithographers struck additional views of many individual cities, leading to the current tally of 4,480 known urban lithographs. Reps, *Views and Viewmakers*, 3, 546.

51. Hans Bergmann, "Panoramas of New York, 1845–1860," *Prospects* 10 (1985): 119–37; Peter B. Hales, *Silver Cities: The Photography of American Urbanization, 1839–1915* (Philadelphia: Temple University Press, 1984).

52. The urban historian John Reps concurred with the nineteenth-century subscribers that the factual information depicted in the urban lithographs is generally reliable. Reps, *Views and Viewmakers*, 67–72.

53. The cultural historian Hans Bergmann explains, "the panoramic illustration or narrative is a cultural production, a creation of the ideology of its time and place. As such it tells us more about the middle-class *ideas* of the city than it does about the actual city." "New York Panoramas," 120. As the historian John

A. Kouwenhoven warned, "a picture of something is not the image itself but somebody's way of looking at it." *The Columbia Historical Portrait of New York: an Essay in Graphic History* (New York: Harper and Row, 1972), 11. For analysis of photography in specific, see Hales, *Silver Cities;* Alan Trachtenberg, *Reading American Photographs: Images as History, Matthew Brady to Walker Evans* (New York: Hill and Wang, 1990); Cervin Robinson and Joel Herschman, *Architecture Transformed: A History of the Photography of Buildings from 1839 to the Present* (New York: Cambridge University Press, 1987); Richard N. Masteller, "Western Views in Eastern Parlors: The Contribution of the Stereograph Photographer to the Conquest of the West," *Prospects* 6 (1981): 56; Malcolm G. Lewis, "Rhetoric of the Western Interior: Modes of Environmental Description in American Promotional Literature of the Nineteenth Century," in *The Iconography of Landscape,* ed. Denis Cosgrove and Stephen Daniels (New York: Cambridge University Press, 1988), 179–93. For analysis of panoramas and bird's-eye prints in specific, see, Reps, *Views and Viewmakers;* Bergmann, "New York Panoramas"; Renzo Dubbini, "Metropolitan Panoramas," *Ottagono* 93 (December 1989): 97–109; Renzo Dubbini, "Views and Panorama: Representation of Landscapes and Towns," *Lotus International* (1986): 98–111. For the western landscape perspective, see Angela L. Miller, *Empire of the Eye: Landscape Representation and American Cultural Politics, 1825–1875* (Ithaca: Cornell University Press, 1993)

54. *Daily America,* 24 November 1847, quoted in Reps, *Views and Viewmakers,* 61.

55. Reps, *Views and Viewmakers,* 65.

56. "Rochester Letterhead," published by Charles Magnus, New York, New York State Archives. The strong journalistic quality of the views made the objectivity of the reporting seem beyond reproach, an effect that lured some unfortunate settlers to boondoggles in the Midwest. Helen Comstock, *American Lithographs of the Nineteenth Century* (New York: M. Barrows, 1950), 14; James Thomas Flexner, *History of American Painting,* vol. 3: *That Wilder Image* (New York: Dover Publications, 1970), 206–7. See also Reps, *Views and Viewmakers,* 17, 39–44, 45–52, 56, 59, 61–62.

57. Marx, *Machine in the Garden,* 226. The dialogue on nature and urbanization has been well studied. See also, Machor, *Pastoral Cities;* White and White, *Intellectual versus the City;* Bender, *Towards an Urban Vision;* Lees, *Cities Perceived;* Cowan, *Cities of the West;* Hales, *Silver Cities.*

58. "A Stroll about Syracuse," *Religious Recorder,* 6 August 1846, OHA Clippings file.

59. Scrantom Diary, 7 August 1831.

60. Hunt, "Good Merchant," *Worth and Wealth,* 57.

61. Stone, "From New York to Niagara," *Buffalo Historical Society Publications,* 233–34.

62. Lieber, *Letters to a Gentlemen in Germany,* 287.

Index

abolitionists, 118–19
aesthetics: beauty, 56, 62–70, 78, 80–81, 86, 107; practice of, 66–70; sublime, 55–56, 57, 59
African Americans, businesses of, 117–19
Agassiz, Louis, 136
agriculture: farmers, 72, 86, 114, 146; improvements to, 155; market system and, 12–15, 163n39, 163n43; small lots and, 36–38
Albany, New York, 27
alcohol, public consumption of, 92–95, 104–5
Allan, Ebenezer, 26
American Journal of Science and Arts, critique in, 67–68
American Magazine of Useful Knowledge, 130
American Scenery (Willis), 56
antiurban rhetoric, 9–10
aqueduct: enlargement of, 140–42; entry to Rochester, 60; original, 43, 54–58
arcades, 125, 129–30. See also Reynolds Arcade
architecture: of arcades, 130; of Athenaeum, 128; of business rows, 78–79; difference between commercial and civic, 52–53; double-fronted buildings, 81–83; higher standards for, 64–65; of industrial district, 53–54; inns as landmark pieces of, 71; at midcentury, 125–26; of milling, 55; as "political technology," 89; of railroad stations, 145; regulations regarding, 84–86; as tool of enticement, 68–69; uniformity and, 83; of weighlocks, 143–44
artist colony, 133–34
assembly halls, 119–20. See also Corinthian Hall
Athenaeum: Bugle Alley and, 136; as civic space, 132; construction of, 127–29; death of W. Reynolds and, 138; as social reclamation project, 134–36; tenants of, 134. See also Corinthian Hall
A. T. Stewart's, 114, 126, 151
Auburn, New York, 19, 144
awnings, 86

Baily, Francis, 27
Baltimore, Maryland, 29, 159n14
banks, 30, 73
Barber, John, 64, 83
Bartlett, Reverend, 138
Bartlett, W. H., 56
Batavia, New York, 19, 47
Beach, Ebenezer, 140
Beecher, Catherine, 69–70
Bethel Church, 95
bird's-eye views, 152–56
Blount, Joseph, 10
Bogardus Inn, 23, 44, 71, 72. See also Mansion House
Boston, Massachusetts, 29, 159n14
bourgeois urbanism, 17, 134–36

Bradley, Lewis, 152, 154
Bremer, Frederika von, 57, 129
Brown, Francis and Matthew, 48
Buckingham, James Silk, 50, 53, 57
Buel, David, 70–71, 76, 137
Buffalo, New York, 19, 29
Buffalo state road, 25, 43. *See also* Buffalo Street
Buffalo Street: as commercial street, 44, 45, 46; dry goods and, 126; B. Hall sketch of, 42–43; N. Rochester and, 25; streetscape of, 83–84. *See also* Reynolds Arcade
Buffalo Street bridge, 44, 124
Bugle Alley, 125, 128–29, 134–35, 136. *See also* Works Street
building city: first stage of, 41–42, 63; pace of, 156; second phase of, 64–65
Burke, Fitzsimmon, Hone & Company dry goods, 113–14
Burns Building, 126
business blocks, 120–22. *See also* Pike Block
business rows: in Rochester, 80–81; in Syracuse, 77–80

canal. *See* Erie Canal
canallers, 94–95
Canandaigua, New York, 19, 47, 144
capitalism, 15
Carroll, Charles: on flood, 54; mill seats and, 31; on property investment, 26; N. Rochester and, 5, 21, 40, 160n16
Carroll Street, 25. *See also* Exchange Street
Carthage, New York, 19, 25
Case & Mann's dry goods, 109–10
cash, 106, 114
cellar establishments, 109–10
Center House, 50, 73
charitable work, 96–97
Charlotte, New York, 19, 25
Cheney, Timothy, 52
Chicago, Illinois, 21, 28
Chicken Row, 92–93, 95
Child, Jonathan, 32, 58, 95, 143

Childs, Timothy, 61
Child's Basin, 43, 60, 142, 143
Chumasero, John: *Life in Rochester*, 99, 109, 110, 114–15; *Mysteries of the Rochester*, 110
Cicero, New York, 19
Circleville, Ohio, 30
city plans: commercial interests and, 20; compactness of, 36–39; elements of, 22; grid platting and, 26–28; intentions of, 39–40; road building and, 13–14; of Rochester, 30–33; sorting and, 28–36; of Syracuse, 30; transportation and, 23; uniformity and, 83–86
cityscape, 8–9, 28–29
civic district: in bird's-eye views, 156; commingling commercial space and, 132; courthouse and, 46–47; genteel view of, 67–68; of Rochester, 35–36, 43, 47–49; subordination of, 46; of Syracuse, 36, 49–53
civilization, city as agent for improving, 12, 15–16, 155
class: alcohol and, 105; in New York, 17; privilege of shoppers by, 113–15; sorting and, 3, 32–33. *See also* merchant class; middle class
Clinton, DeWitt: on erecting villages, 19; Erie Canal and, 13, 60; Halls and, 175n51; Native Americans and, 104, 185–86n75; on Rochester, 62
Clinton House, 73, 92
Clinton Square: as commercial district, 33, 45, 65; double fronts on, 81–83; hotels refine, 73–77; platting and, 24; as symbol of public life, 20–21
clustering as self-fulfilling prophecy, 45
commercial district: in bird's-eye views, 155–56; density of, 45–46; double fronts in, 81–83; as epitome of city, 66; moving merchants and, 152; public space within, 131–36; refinements to, 66–69, 150; regulations regarding, 84–86; of Rochester, 30–31, 32–34, 42–43, 44–46, 124–25; shop

INDEX

houses and, 3, 32, 78; social sorting in, 88–89, 135–36; subordination of, 67–68; as symbolic space, 65–66; of Syracuse, 20–21, 24, 44; train station and, 145–46. *See also* business rows; social construction of space; social sorting in commercial buildings
commercial urbanism, 37–38
communication, 130–31
compactness of city, 36–39, 42–44
consumer revolution, 88
Cooper, William, 37–38, 39
Cooperstown, New York, 37
Corinthian Hall, 128, 132, 134, 136
Cornell, Silas, map of, 49
county seat: Rochester as, 35; Syracuse as, 36; transportation and status as, 25
courthouse: in Rochester, 47–49; in Syracuse, 49–53
courts and economic growth, 11
crime, 135
"crystal palaces," 129
cultural improvement, 14–16, 137–38

daguerrean studios, 134
Dana, Charles, *United States Illustrated*, 56
Davies, Phebe, 99
demographics of New York State, 1
Detroit, Michigan, 29
Dewey's News Room, 153
DeWitt, Benjamin, 14
Dickens, Charles, 17
Dillaye, Henry, 79, 146
disorderly conduct in public, 97–98
Doolittle, Araminta, 99
double-fronted buildings, 81–83, 141
Douglass, Frederick, 118, 136
Drake, Daniel, 27
dry goods rows, 110–11, 122, 151
dry goods stores: Buffalo Street and, 126; Burke, Fitzsimmon, Hone & Company, 113–14; Case & Mann's, 109–10; Empire Buildings, 111; Emporium Buildings, 80, 114; Lynch Brothers, 152; McCarthy's, 151; Wilder & Gorton, 113
Duttenhofer, Adolphus, 43–44, 55–56

Eagle Tavern and Hotel: Bank of Monroe and, 73; history of, 34; refurbishments to, 76, 92, 101; sketch made from vantage point of, 42
East Rochester, 19, 47, 48
economy: financial panic, 140, 141; improvement and, 9–14, 139; merchant class and, 17, 61–62, 65; political, 4–5; urban views and, 154–55
election day, 51, 92
Elliot, Daniel, 145
Ely, Hervey, 141, 177n4
Emerson, Ralph Waldo, 27, 136
Empire Buildings and dry goods, 111
Empire House, 76, 111. *See also* Mansion House; Vorhees House
Emporium Buildings and dry goods, 80, 114
Ensworth House, 34, 76
enterprise, 10–11
entrepreneurship, 7, 16–17, 65
Erie Canal: Broad Street Aqueduct Bridge, ix; celebrations, 60–63; cultural improvement and, 14–16; enlargement of, 139–44; Erie Boulevard, ix; facades on, 82; Genesee River and, 43; Hawthorne on, 6; locating, ix; mercantile network and, 12–14, 24; opening of, 1, 60–63; proposal for, 13; real estate prices and, 27; Rochester and, 57–58; swimming in, 98; Syracuse and, 6; ticket office for, 69; toasts at opening of, 12, 13, 14, 61, 62; towpaths, 81, 90, 92–95, 106; weighlocks, 92, 143–44
Esperanza, New York, 29
Exchange Street, 58, 142. *See also* Carroll Street

Farrell, Mary, 100, 117
Female Charitable Society, 96–97, 136, 194n45

Fifty-first Regimental Armory, 150
Fisher, Alvin, *Remnant of the Tribe*, 106
Fitzhugh, William, 5, 21, 26, 160n16
flood in Rochester, 54, 124
Forman, Joshua: boosterism of, 36–37; debt of, 168n23; hotel and, 73–74; on improvement, 12; life of, 160n17; as nonplanner, 7; platting by, 23, 24, 32, 36; salt manufacture and, 58–59; Syracuse and, 2, 6; toast at opening of Erie Canal, 14; Walton Tract and, 27
Forman, Owen, 7, 24
Four Corners: coaxing commerce into, 33; compactness of, 42; as heart of city, 20–21, 30–31; land values in, 34, 125; in 1920s, ix. *See also* Reynolds Arcade
Fowler, John, 26, 57
Fowler, Orson Squire, 136
Frankfort, New York, 19, 21–22, 47, 142
Franklin Buildings, 83
Freeman, Lois, 114
frontier: cities along, 2, 4; mercantile chain and, 53, 130–31; planning and, 39–40; settling of, 1, 10, 11–15, 162n29; sorting and, 155–56
functional sorting in cities, 28–29

gable ends of buildings, 46
Geddes, New York, 19
Genesee County, 47
Genesee Farmer (magazine), 69, 130
Genesee River: Erie Canal and, 43; falls of, 55–57; flood of, 54, 124; Middle Falls, 55, 56; in Rochester, ix; settlements along, 19; Upper Falls, 16, 24, 54
Geneva, New York, 19, 144
gentility: of civic district, 67–68; of hotels, 91–92, 101–2; of merchant class, 1, 17, 39, 87–89; of public space, 131–36; of Reynolds Arcade, 127, 129, 131–32; of store interiors, 111–12; women and, 96, 99–100
gentrification, 87

Globe Block, 146
Globe Buildings, 124
Gould, Lindley, 119
Granger, Amos, 74–75
Granger Block, 150
Greek Revival style, 78–79, 143, 145
Greeley, Horace, 136
grid platting, 22, 23, 26–28
groceries, 92–94

Hagerstown, Maryland, 21
Hall, Basil: Clinton and, 175n51; on improvement, 11; on Rochester, 41–42, 63; on salt manufacture, 59–60; sketch of Buffalo Street by, 42–43, 83–84
Hall, James, 102–3
Hall, Margaret, 41, 63, 175n51
Hand, Marcus C., 22, 79
Hanford's Landing, New York, 19
Hanover Square: business row in, 78–79; fire in, 77, 83; as heart of city, 20–21, 24
Hawthorne, Nathaniel, 6, 72, 86
Hill, John W., 152, 153
Holmes, Oliver Wendell, 136
Hopkins, Alice, 99
hotels: Bogardus Inn, 23, 44, 71, 72; Center House, 50, 73; Clinton House, 73, 92; day use of, 72–73; definition of, 179n34; Empire House, 76, 111; Ensworth House, 34, 76; gentility and, 91–92, 101–2; interiors of, 91–92; Mansion House, 44, 73, 76, 91; Monroe House, 73; porches of, 90–91; privacy for women in, 101–2; in Rochester, 76–77; Rochester House, 73, 92, 142–43; as stimulating development, 71–72; in Syracuse, 73–76; Syracuse House, 73–75, 76, 91, 101–2, 118; United States Hotel, 72–73; Vorhees House, 51. *See also* Eagle Tavern and Hotel
Howe, Henry, 64, 83
Howison, John, 72
Hunt, Freeman: on commerce, 155; on

INDEX 205

men in business world, 116; on retail business, 113; on River Street in Troy, 29; to shopkeepers, 152; on urban merchant, 16, 17
Huntington, Alcesta, 100, 101, 134
"Hymn for Female Penitents," 99

illustrations: bird's-eye views, 152–56; in gazetteer of New York State, 64, 77–78, 83–84, 103–4; of Rochester, 42–44, 83–84; of salt manufacture, 59
improvement: city as vehicle of, 10–11, 156; falls and, 55–57; Native Americans and, 107; overview of, 11–12; social and cultural, 14–16; social and economic, 139
industrial district: architecture of, 53–54; planning, 31–32; of Rochester, 43–44, 54–58, 150; of Syracuse, 58–60, 149. *See also* milling district of Rochester
industry, 10–11
information links, 130–31
invisibility of women, 98–99
Italianate style, 79, 121, 128

Jackson, Bennett, 118
Jacobs, Harriet, 118–19
James, William, 74
Jarves, James Jackson, 71, 76, 137
Jefferson, Thomas, on cities, 9
Jefferson Square, 149
Jeffersonville, Indiana, 30
Jerome, Larry, 113
Johnson, Elisha, 19, 48, 124
Journal Building, 82–83

Kemble, Fanny, 136
Kempshall, Mrs., 97

landscape: beautification of, 69; canal enlargement and, 141; commercialization of, 8; developer's legacy on, 39–40; eradication of Native American, 14–15; improvement and, 11–12; urban views and, 154
language, indecent, in public, 97–98

Larned Building, 150
Leavenworth, E. W., 69, 149
Leighton, James and John, 183n31
Life in Rochester (Chumasero), 99, 109, 110, 114–15
lighting, night, 112
Lind, Jenny, 136
lithographs, 152–56
Littles, Sarah, 100
Liverpool, New York, 19
loitering in public: Native Americans and, 106; white males and, 90–91; women and, 101
London, England, 28, 129
Lynch Brothers dry goods, 152

Magnus, Charles, 153
Main Street, 25, 124
Manifest Destiny, 10
Manlius, New York, 19, 164n48
Mansion House, 44, 73, 76, 91. *See also* Bogardus Inn; Empire House
Marble Palace, 126
Masonic Hall, 126
Matthews, Vincent, 61
McCallum, D. C., 126
McCarthy, Dennis, 151–52
McCarthy, Thomas, 151
McCarthy's dry goods, 151
Mechanics Association, 127, 128
Melville, Herman, 94
men. *See* white males
mercantile city, image of, 154–55
mercantile colleges, 116
mercantile ethos, 1, 12–14, 16–17. *See also* merchant class
mercantile triangle, 28–29
merchant class: architectural refinement and, 70–71, 87–88; as carrier of culture and morality, 16–17, 137–38; gentility and, 1, 17, 39, 87–89; leadership of, 61–62, 65; as patron of city, 70–71, 137; regulations of regarding architecture, 84–86; social codes of, 87, 88–89; social reclamation project of, 134–36. *See also* middle class

merchants. *See* merchant class
middle class: architectural standards and, 65; compactness and, 39; domesticity and, 96; rise of, 87; sobriety and, 92–95. *See also* gentility; merchant class
millinery shops, 117
milling district of Rochester, 31–32, 53–60, 140. *See also* industrial district
millpond in Syracuse, 149–50
Mill Street, 58
Monroe County, ix, 35, 43
Monroe County courthouse, 47–49
Monroe Hall, 120
Monroe House, 73
Montez, Lola, 136
moral imperatives: city as carrier of economic productivity and civilization, 10, 155; improvement, 11–12, 15; merchant and, 137–38; temperance, 93, 94–95, 105
movement, continuous: of Native Americans, 105–6; of women on street, 99–100
multiplier effect, 37
Myers, J. C., 10, 14, 107
Mysteries of the Rochester (Chumasero), 110

Native Americans: alcohol and, 104–5; cultural differences of, 102–3; eradication of, 14–15; exclusion of, 107–8; in gazetteer, 103–4; groceries and, 93; money and, 106–7; movement of, 105–6; public space and, 102; at Seneca ceremonial site, 103; white views of, 104
New York City: Crystal Palace, 129; grid plan for, 27–28, 29, 30; land uses in, 158n7; Marble Palace, 126; new cities' similarity to, 29, 65, 66, 78, 79; population of, 159n14; trade connections to, 4, 13, 14, 60

Odd Fellows Hall, 120
100–Acre Tract, 26, 31, 38, 40
one-price system, 114–15

Onondaga County, 36, 49, 52–53
Onondaga County courthouse, 49–53
Onondaga Hill, 49, 50
Onondaga Hollow, 19, 36, 49
Onondaga Lake, 58
Onondaga Salt Company, 59
Onondaga tribe, 187–88n103
Ontario County, 47
open space, 30
order, 66, 83–89. *See also* grid platting
O'Reilly, Henry, *Sketches of Rochester*, 13, 103
Orion Club, 92
Oswego, New York, 160n15

Panorama of Professions and Trades (Hazen), 115–16, 117
Parker, Jenny Marsh, 58, 130
Peck, Everard, 140
Pennsylvania town model, 21
Philadelphia, Pennsylvania, 29, 157–58n7, 159n14
Phinney's Museum, 107
Phoenix Buildings, 77, 79, 83
Pike Block, 79, 120, 146
platting, 22, 26–28
Pompey, New York, 19
post office, 125, 130, 147
Powell, Harriet, 118
Powers Block, 34
Providence, Rhode Island, 125, 130
public space, commercially conditioned, 20–22, 131–36
public space, social sorting in: bourgeois females, 96–102; Native Americans, 102–8; overview of, 89–90; white males, 90–95
Public Square Alley, 42
public watch, 135
public woman, 99

railroad: Binghampton, 150; canal and, 140; congestion, smell, and sound of, 147–48; mercantile network and, 13–14; New York Central, 142, 144; relocation of station for, 148–50; in

INDEX 207

Rochester, 142; stations for, 145–47; in Syracuse, 144–45; Syracuse and Utica, 148
Rann, Augusta, 93, 99–100, 101, 145
Raumner, Frederick von, 56
reading, 101
real estate prices: canal expansion and, 141; competition and, 38; Erie Canal and, 27; railroad and, 146; sorting and, 33–34
Red Mill, 58
Renwick, James, 121
residential districts, 3, 32–33
retail business: class and privilege in, 113–15; double-fronted buildings and, 81; dry goods rows, 110–11, 122, 151; as gendered, 109–10; parlors in, 111–12; shop floor, 113; wholesale business and, 45, 81, 148–49; windows in, 112. *See also* dry goods stores
Reynolds, Abelard: Arcade and, 124–25; bust of, 138; finances of, 177n4; lot bought by, 34; praise for, 137; public watch and, 135. *See also* Reynolds Arcade
Reynolds, William: Arcade and, 126–27, 130; Athenaeum and, 132; Douglass on, 119; mourning for, 136–38
Reynolds Arcade: architectural refinements to, 126–27; architecture of, 130; Athenaeum and, 127–28; business of business and, 130–31; construction and remodeling of, 125; cost of, 127; death of W. Reynolds and, 138; description of, 123, 138; genteel tableau of, 127, 129, 131–32; patrons of, 133; "The Rochester Arcade Quick Step," 129; as social reclamation project, 134–36; spatial sorting within, 133–34; as urban promotion, 129–31
road building, 13–14, 22, 33
Robber's Row, 92–93, 95
Rochester, Nathaniel: bridgehead and, 24–25; Buffalo state road and, 25;

Carroll and, 40; compactness principle and, 38–39; contracts of, 38–39; county seat and, 47, 48; Erie Canal and, 13–14, 176n69; in Hagerstown, 21; life and career of, 160n16; as nonplanner, 7; 100–Acre Tract and, 26; platting by, 19, 23, 30–31, 32, 35–36; prices for lots sold by, 33–34; as project overseer, 5; Reynolds and, 124
Rochester, New York: city plan of, 20, 30–33; commercial development and, 16; courthouse in, 47–49; east and west banks of, 124; founding of, 5; illustrations of, 42–44, 154–55; legal incorporation of, 32; opening of Erie Canal and, 60–63; platting of, 19, 23; population of, 64, 152, 159n15, 190n144; Trollope on, 4. *See also* Four Corners
Rochester Female Academy, 99
Rochester House, 73, 92, 142–43
Rochester Industrial School, 142
Rochester Music Store, 111
Rolph, Thomas, 57
Russell, Archimedes, 151

Sabbatarians, 94–95
Salina, New York: annexation of, 51; commercial district of, 151; as county seat, 49, 50; fights between Syracuse and, 50–51; founding of, 19; salt springs and, 58–59; Washington Square, 21–22
Salina Street bridge, 90
salt manufacture, 58–60, 144–45, 149
sawmill, 23–24
scape, definition of, 8–9
Scrantom, Edwin: on Bugle Alley, 128–29, 135; as clerk, 116; on commercial architecture, 80; on election day drunkenness, 92; music published by, 76–77; on Native Americans, 102, 105, 107–8; on Rochester, 63, 155; on Rochester House, 142–43; on view outside store, 90

Searll, Henry, 128
Second Great Awakening, 15, 17, 88
Seneca Turnpike, 71
Seward, Austin, 118
shipping center in Rochester, 57–58, 81
shop houses, 3, 32, 78
Sibley, Elisha, 106
signs, 85
sitting in public: men and, 90–91; women and, 101, 134
Sketches of Rochester (O'Reilly), 13
Skinner, Henry, 34
Smith, E. L., 134
Smith, Vivus, 146
Smith Block, 84
smoking, 91–92
sobriety and middle class, 92–95
social construction of space: for bourgeois females, 96–102; for Native Americans, 102–8; overview of, 7, 89–90; for white males, 90–95
social improvement, 14–16
social production of space, 7
social reclamation project in Rochester, 134–36
social sorting in commercial buildings: cellars, 109–10; overview of, 108; street level, 110–15, 133, 134; top floor, 119–22; upper floors, 115–19, 133–34; Yates Block, 120–22
Sophia Street, 141
sorting: cityscape and, 9; definition of, 1; good urban form and, 18; planning for, 28–36; Syracuse as example of, 2
South Salina Street, 146
Spafford, Horatio Gates, *New York State Gazetteer*, 26
spatial culture, 6–8
Spencer, John, 61
squares, civic, 20, 21
Steele, Eliza, 70, 75
St. Luke's Episcopal Church, 49
Stone, Enos, 7
Stone, William Leete: Bogardus Inn and, 72; on Rochester, 62–63, 156; on Salina, 51; on Syracuse, 1–2, 18

store interiors, gentility of, 111–12
streetscape, 41, 83–84
street walker, 99
Strong, Augustus, 92, 116
subordination of civic to commercial activities, 35–36
Syracuse: bird's-eye view of, 154, 155; business directory for, 66; central heart of, 20–21; city plan of, 20, 30; commercial development and, 16; courthouse in, 49–53; fights between Salina and, 50–51; founding of, 5–6; platting of, 23; population of, 64, 152, 159n15, 190n144; railroad in, 145–50; riot in, 51; Stone on, 2–3
Syracuse Book Store, 101
Syracuse Company: architectural improvements of, 77; civic area and, 49–50; influence of, 75–76; lot sales and, 38; Onondaga millpond property of, 149; Syracuse House and, 74, 75; Walton Tract and, 27. *See also* Forman, Joshua; Townsend, Isaiah; Townsend, John
Syracuse House: first building of, 73–74; porch of, 91; second building of, 74–75; slave rescue and, 118; Vandewater on, 75, 76; women's parlors in, 101–2
Syracuse Salt Company, 59

Tallmadge, James, 60, 62
tavern, 125, 179n34
telegraph office, 130, 131, 147
temperance reformers, 93, 94, 105, 120
Tourist Guide (Vandewater): on business rows, 77–78; on Rochester, 65, 124; on Syracuse House, 75, 76
tourists: arcades and, 129; to mills and falls, 57; salt manufacture and, 59; travel writing and, 26
Townsend, Isaiah, 58–59
Townsend, John, 38, 77, 149
Townsend Block, 77
towpaths, 81, 90, 92–95, 106
transportation: city plan and, 23; com-

INDEX 209

mercial posturing of cities and, 22; county seat status and, 25; improvement and, 12–14; in maturing city, 139; in Rochester, 24–25; settlement and, 25–26; in Syracuse, 23–24. *See also specific modes*
travel writing, 26
Trollope, Anthony, 72
Trollope, Fanny, 4, 57
Troy, New York, 29, 70–71, 196n26
True Politeness for Ladies, 98, 99
turnpikes, 4, 5, 19, 71

underground railroad, 118
uniformity, visual, as goal, 83–86, 127
United States Hotel, 72–73
United States Illustrated (Dana), 56
urban promotion, 71–72, 129–31, 152–54
Utica, New York, 19

Vandewater, R. J., tourist guide of. *See Tourist Guide* (Vandewater)
Van Every, Mrs. C. C., 117
vernacular urbanism, 1, 5, 6–8
Verplanck, Johnston, 39
vice, 99, 135. *See also* alcohol, public consumption of
visibility and commercial buildings, 112–13
Vorhees, Colonel, 51–52, 73, 76, 111
Vorhees House, 51. *See also* Empire House; Mansion House

walking city model, 3–4
Walton, Abraham, 5, 23, 71
Walton Tract, 27
Wanamaker's, 114
Warner, Sam Bass, Jr., 3
Washington, D.C., 29
Washington Block, 79–80, 146

Webster, Daniel, 136
Webster, Noah, on improvement, 11
weighlocks, 24, 143–44
Western Union Telegraph office, 130, 131
Wharton, Thomas, 106
White, Horatio Nelson, 52, 150
Whitefield, Edwin, 153
white males: Arcade and, 133; city building and, 4–5; in commercial district, 90–95; as merchants, self-interest of, 16–17; Native American landscape and, 14–15; opening of Erie Canal and, 61–62; settlement of New York interior by, 1; upper floors and, 115–16
White Memorial Building, 150
The Whole Duty of Woman, 96, 98
wholesale business: cellars and, 109; railroad and, 150; retail business and, 45, 81, 148–49
Wilbur, Andrew, 119
Wilder & Gorton dry goods, 113
Wilkinson, John: as postmaster, 147; real estate speculation by, 146; south side and, 45; as surveyor, 7, 24; Syracuse name and, 6
Williams & Hutchinson, 152
Willis, Nathaniel, *American Scenery*, 56
women: commercial buildings and, 111–12; employment of, 116–17; gentility and, 96, 99–100; Native American, 104–5; public space and, 96–102; as shop clerks, 113
Woodhull, Victoria, 136
workhouse in Syracuse, 97
Works Street, 125, 129, 134. *See also* Bugle Alley
Wright, Benjamin, 43

Yates Block, 120–22

Illustration Credits

Fig. 1. John W. Barber and Henry Howe, *Historical Collections of the State of New York; Containing a General Collection of the Most Interesting Facts, Traditions, Biographical Sketches, Anecdotes, Etc.* (New York: S. Tuttle, 1841).
Fig. 2. Based on 1811 plat by Nathaniel Rochester, redrawn by Kai Gutschow.
Fig. 3. Kai Gutschow.
Fig. 4. Based on 1819 plat by Owen Forman and John Wilkinson, redrawn by Kai Gutschow.
Fig. 5. Kai Gutschow.
Fig. 6. Field notes of Lemuel Foster, *New York State Laws of 1815*, chap. 224.
Fig. 7. Carroll E. Smith, *Pioneer Times in the Onondaga Country* (Syracuse: C. W. Bardeen, 1904), 240.
Fig. 8. Basil Hall, *Forty Etchings, from Sketches Made with Camera Lucida in North America, in 1827 and 1828* (London: Cadell & Co., 1830), x. From the "Rochester Images" collection, courtesy of the Local History Division, Central Library of Rochester and Monroe County.
Fig. 9. A. Duttenhofer, *Bereisung der Vereinigten Staaten von Nordamerika* (Stuttgart: C. W. Löflund, 1835). From the collection of the Rochester City Hall Photo Lab.
Fig. 10. "Rochester Images" collection, courtesy of the Local History Division, Central Library of Rochester and Monroe County.
Fig. 11. Onondaga Historical Association Museum and Research Center, Syracuse, N.Y.
Fig. 12. Henry O'Reilly, *Settlement in the West: Sketches of Rochester; with Incidental Notices of Western New York* (Rochester: W. Alling, 1838), following 360.
Fig. 13. O'Reilly, *Settlement in the West*, following 376.
Fig. 14. 1838 view of the Upper Falls of the Genesee River, by W. H. Bartlett.
Fig. 15. J. H. French, *Gazetteer of the State of New York . . . and a Complete History and Description of Every County, City, Town, Village, and Locality* (Syracuse: R. Pearsall Smith, 1860), following 480.
Fig. 16. Barber and Howe, *Historical Collections of the State of New York*, facing 269.
Fig. 17. Barber and Howe, *Historical Collections of the State of New York*, 395.

Fig. 18. "Architecture in the United States," *American Journal of Science and Arts* 17 (January 1830): 260.
Fig. 19. Marcus Christian Hand, *From a Forest to a City: Personal Reminiscences of Syracuse* (Syracuse: Masters & Stone, 1889), facing 32.
Fig. 20. Hand, *From a Forest to a City*, facing 34.
Fig. 21. "Syracuse, New York," *Ballou's Pictorial Drawing Room Companion* 11:5, 2 August 1856, 72.
Fig. 22. Courtesy of the Erie Canal Museum, Syracuse, New York.
Fig. 23. Onondaga Historical Association Museum and Research Center, Syracuse, N.Y.
Fig. 24. Courtesy of the Erie Canal Museum, Syracuse, New York.
Fig. 25. Onondaga Historical Association Museum and Research Center, Syracuse, N.Y.
Fig. 26. Robert N. Dennis Collection of Stereoscopic Views, Miriam and Ira D. Wallach Division of Art, Prints and Photographs, The New York Public Library, Astor, Lenox and Tilden Foundations.
Fig. 27. Gurney S. Strong, *Early Landmarks of Syracuse* (Syracuse: Times Publishing, 1894), 271.
Fig. 28. H. J. Sutherland, *The City of Syracuse and Its Resources Illustrated* (Syracuse: Syracuse News Publishing, 1893), pl. 34.
Fig. 29. H. P. Smith, *Syracuse and Its Surroundings* (Syracuse: Hamilton Child, 1878), pl. 62.
Fig. 30. Image courtesy of the Department of Rare Books and Special Collections, University of Rochester Library.
Fig. 31. O'Reilly, *Settlement in the West*, following 376.
Fig. 32. "Rochester Images" collection, courtesy of the Local History Division, Central Library of Rochester and Monroe County.
Fig. 33. American Sunday School Union's *Common Sights in Town and Country*. Photo courtesy of The Philadelphia Print Shop, Ltd.
Fig. 34. *Rochester City Directory, 1853–54* (Rochester: D. M. Dewey, 1853), 21. Image courtesy of the Rare Books and Special Collections, University of Rochester Library.
Fig. 35. *Boyd's Rochester Directory with a Business Directory* (Rochester: Boyd, 1863), 171. Image courtesy of the Rare Books and Special Collections, University of Rochester Library.
Fig. 36. Edward Hazen, *The Panorama of Professions and Trades; or Every Man's Book* (Philadelphia: Uriah Hunt, 1837), 195.
Fig. 37. *An Abridged Specimen of Printing Types Made at Bruce's New-York Type-Foundry* (New York: George Bruce's Son, 1869), 151.
Fig. 38. Image courtesy of the Rare Books and Special Collections, University of Rochester Library.
Fig. 39. Hazen, *Panorama of Professions and Trades*, 61.

Fig. 40. Hazen, *Panorama of Professions and Trades*, 175.
Fig. 41. *Troy City Directory for 1865* (Troy, 1865), 188.
Fig. 42. H. Perry Smith, *The Royal Road to Wealth, an Illustrated History of the Successful Business Houses of Syracuse* (Syracuse: Vanarsdale, 1873), following 10.
Fig. 43. Smith, *Royal Road to Wealth*, following 12.
Fig. 44. Smith, *Royal Road to Wealth*, following 12–15.
Fig. 45. Kai Gutschow.
Fig. 46. Image courtesy of the Department of Rare Books and Special Collections, University of Rochester Library.
Fig. 47. "The Agricultural Society Fair," *Illustrated American News, Supplement*, 1851. Image courtesy of the Department of Rare Books and Special Collections, University of Rochester Library.
Fig. 48. Derived from *Rochester City Directory, 1862–63*, 62–64; and *Illustrated Guide to Reynolds Arcade, with History of the Arcade from 1828, Plans of Each Floor, Number of Each Office, Names and Business of Each Occupant* (Rochester: Harris & Foxwell, 1885), 10. Drawn by Kai Gutschow.
Fig. 49. "Rochester Images" collection, courtesy of the Local History Division, Central Library of Rochester and Monroe County.
Fig. 50. "The Agricultural Society Fair," *Illustrated American News, Supplement*, 1851. From the "Rochester Images" collection, courtesy of the Local History Division, Central Library of Rochester and Monroe County.
Fig. 51. *Illustrated Guide to Reynolds Arcade, with History of the Arcade from 1828, Plans of Each Floor, Number of Each Office, Names and Business of Each Occupant* (Rochester: Harris & Foxwell, 1885), 10.
Fig. 52. "Syracuse, New York," *Ballou's Pictorial Drawing Room Companion* 11:5, 2 August 1856, 72.
Fig. 53. Smith, *Syracuse and Its Surroundings*, pl. 25.
Fig. 54. Onondaga Historical Association Museum and Research Center, Syracuse, N.Y.
Fig. 55. Lewis Bradley, December 1852. N. Phelps Stokes Collection, Miriam and Ira D. Wallach Division of Art, Prints and Photographs, The New York Public Library, Astor, Lenox and Tilden Foundations.
Fig. 56. "Bird's-eye View of Syracuse, New York, 1874." Courtesy of the Erie Canal Museum, Syracuse, New York.
Fig. 57. John William Hill, 1854. N. Phelps Stokes Collection, Miriam and Ira D. Wallach Division of Art, Prints and Photographs, The New York Public Library, Astor, Lenox and Tilden Foundations.

About the Author

Diane Shaw was born in Ann Arbor, Michigan, and raised in Amherst, Massachusetts. She received her doctorate from the University of California at Berkeley in 1998. Her current research and writings focus on the cultural aspects of the American built environment, emphasizing the vernacular and ordinary patterns of building within their social and urban context. Her articles have appeared in *Perspectives in Vernacular Architecture V,* the *Journal of Medieval and Early Modern Studies,* and *Washington History.* She is an associate professor in the School of Architecture at Carnegie Mellon University and the series editor of *Vernacular Architecture Studies,* which is sponsored by the Vernacular Architecture Forum and published by the University of Tennessee Press.

Related Books in the Series

Petrolia: The Landscape of Pennsylvania's Oil Boom
   Brian C. Black

Historic American Towns along the Atlantic Coast
   Warren Boeschenstein

America's New Downtowns: Revitalization or Reinvention?
   Larry R. Ford

Cities and Buildings: Skyscrapers, Skid Rows, and Suburbs
   Larry R. Ford

Unplanned Suburbs: Toronto's American Tragedy, 1900 to 1950
   Richard Harris

Magnetic Los Angeles: Planning the Twentieth-Century Metropolis
   Greg Hise

Boston's "Changeful Times": Origins of Preservation and Planning in America
   Michael Holleran

Greenbelt, Maryland: A Living Legacy of the New Deal
   Kathy D. Knepper

Manufacturing Montreal: The Making of an Industrial Landscape, 1850 to 1930
   Robert Lewis

Entrepreneurial Vernacular: Developers' Subdivisions in the 1920s
   Carolyn S. Loeb

Suburban Landscapes: Culture and Politics in a New York Metropolitan Community
   Paul H. Mattingly

From Tavern to Courthouse: Architecture and Ritual in American Law, 1658–1860
  Martha J. McNamara

The Sanitary City: Urban Infrastructure in America from Colonial Times to the Present
  Martin V. Melosi

The Birth of City Planning in the United States, 1840–1917
  Jon A. Peterson

America's Original GI Town: Park Forest, Illinois
  Gregory C. Randall

John Nolen and Mariemont: Building a New Town in Ohio
  Millard F. Rogers

Apostle of Taste: Andrew Jackson Downing, 1815–1852
  David Schuyler

Public Markets and Civic Culture in Nineteenth-Century America
  Helen Tangires

The Rough Road to Renaissance: Urban Revitalization in America, 1940–1985
  Jon C. Teaford

Redevelopment and Race: Planning a Finer City in Postwar Detroit
  June Manning Thomas

The City Beautiful Movement
  William H. Wilson

Hamilton Park: A Planned Black Community in Dallas
  William H. Wilson